STANDARD ENVIRONMENT IN SRI LANKAN TEA TRADE

Vindika Lokunarangodage

Indira Wickramasinghe

2018

ISBN 978-955-35840-2-1

Acknowledgements

At the outset of Standard Environment of Sri Lankan Tea Trade, authors gratefully acknowledge the dedicated support of tea manufacturers in Sri Lanka, specially low grown orthodox black tea manufacturers who allowed the use of their facilities for research proposes while evaluating the existing conditions and available systems as well as providing insights of the problem they encounters and their expectations.

It was indeed a pleasure working with the University of Sri Jayawardenapura, who allowed to use their facilities, and human resources for the research purposes of the work.

Authors highly indebted to the Prof. K. K. D. S Ranaweera, for his tireless support, motivation, comments, and evaluation of the manuscript.

The list of tea factories and the persons involved in the discussions regards to the industry were very longer but without their support, this work would have been not possible where author's sincere thanks go to all personnel including tea factory owners, and their staffs who were participated in interviews, discussions and evaluations which further includes smallholding farmers, collectors, managers of companies and factories, auctioneers and auditors as well as quality officers.

Finally, authors acknowledges friends, family and the society including technology that helps converting these gross ideas into a manuscript.

Preface

The book is written around the various standards applied in tea Sri Lankan tea industry but it not only represent the local tea trade. The Sri Lankan tea industry is usually manufacturing far better quality teas compared to the many tea manufacturers in the world. The tea industry in one of the oldest as to the history and in the country it is almost century and half old with the status of most consumed beverage after the water. Most of the factory owners, management staff, consultants etc., argue that there is no any financial gain over achieving a good quality/safety systems due to high cost involved over no additional price gains in the auctions.

Nonetheless, recent events, changes in management and short sighted new generation of factory owners and the politicians are destroying the centuries old, world renowned tea trade in Ceylon. The industry do not intend to develop a better quality/safety assurance systems, instead they are interested to obtain a certificate for the marketing purposes, hence they reject the idea and trying to create additional revenue through various illegal activities. Thus, younger generations who enter the industry learning from them and try to ignore the new realities. On the other hand, the country's major agricultural export is tea with one of the largest work forces where the lowest educated poor marginalized are working.

Hence, book is trying to provide reliable information for factory owners, employees, students, trainees and the workers an opportunity to understand the standard environment and relevant data while implementing, quality, food safety, occupational safety, social and environmental etc., or their basics. Further, book discusses, current trade related issues, and future of the standard environment and recent developments in the trade.

In addition to that, this book is also providing as many as possible food safety management related information, which can be used in combination with ISO 22000:2018 generic model that was recently publish by author. The book provides most of the information required by people who try to develop food safety management systems in the tea industry while providing additional information of relevant other standards in the industry to provide an overall picture, because there is a growing concern over food safety quality, fairness to employees and the environment that teas are grown.

Vindika Lokunarangodage
Indira Wickramasinghe
September, 2018.

Table of Content

1
Introduction

Tea was known as a beverage for at least 5000 years, it was initially considered as a medicine and later grew into a beverage which finally became the most popular beverage in the world or the second most important drink after water. According to the Chinese legends, tea was first discovered in the time of Second Emperor and herbalist, Shennong in 2737 B.C.E, [1] accidently when a dried tea leaf was fallen into the boiling water which is intended for emperor's thirst while he was travelling to another region [2], who drank boil water as a habit. In the time of Han Dynasty (206 BCE – 220 CE), tea was used as a medicine and much later tea became popular as a beverage and people started drinking it in social events at Tang Dynasty around 618 – 2907 CE [1].

In seventeenth centaury, tea was introduced to the west where it became quickly popular due to the flavour and stimulative properties extracted to boiling water [3] because of the cold climatic conditions; hot water drinking was a habit in those societies and due to the poor hygiene of water during this era, boiling of water was common place. In fact, first tea consignments were reached in London in 1652, where there were hygiene threats due to the lots of waterborne diseases as a result of industrialization and pollution of urban water courses. Thus, people used to drink boiled water which has no fresh taste where addition of tea gave it both flavour and taste due to the presence of caffeine [4].

Even though coffee and chocolate also introduced in to the Europe in same era, both of them were unaffordable for the general public until after second world war and other alternative was alcohol which also kill the pathogens, but it was not practical solution where tea became more popular than other beverages [3]. Tea even became a seed for wars, such as the opium war in China (1840 – 1842) which irreversibly changed the entire cultures with positive and negative effects. Today, it is a multimillion dollar industry with over USD 4 billion revenue and growing while employing more than 15 million people globally and providing over four billion tea cups a day all over the world [5]. In addition, current research findings on the health aspects of tea drinking also prove that the first assumptions of drinking tea as a medicine is true all along the history as it assumed by Chinese people without proper evaluations [3].

Tea, *Camellia sinensis* belongs to the family Theaceae (Camelliaceae). The genus Camellia was developed in Asia centered on the Himalayan Mountains. Nevertheless, the two types of Camellia with stimulant properties developed separately. *Camellia sinensis* Var. sinensis on the northern slopes of Himalayan while C. sinensis Var. assamica on the southern region and adjoining plains [6]. The Cambodian type makes third type of tea [7]. Chinese tea bushes (***Camellia sinensis***), the Assam tea tree (***Camellia assamica***) and their hybrids, but Georgian and Sinhalese tea are the most popular. ***Camellia sinensis (L) O.*** Kuntze is the main export crop in Sri Lanka under Ceylon Tea. The commercial planting of tea in Sri Lanka was introduced by a Scotsman, James Taylor in 1867 [8]. Since then the Ceylon Tea has been the world's number one brand for over a century and the local tea industry has made a significant growth over the

time while securing its position in the global market as a leading producer and exporter of high quality black tea.

Accordingly, Sri Lanka is one of the oldest tea producing countries in the world and it commenced the commercial scale tea production in 1867 [8]. They were better known in the world as 'Ceylon Tea' while ranking among the best available teas in international trade [9]. A survey conducted by TRI in 2003 [10] revealed the productivity as 2200 kg per hectare in vegetatively propagated cultivars and the amount was much less in seedling types in estate plantation companies. The productivity levels achieved in Sri Lanka was far below than those of other competing nations such as India, Kenya, Vietnam and Indonesia. Until 2004, Sri Lanka was the leading tea exporter in the world, later Kenya took this place and Vietnam also exceeded the Sri Lankan export volumes. In 2007, Sri Lanka was the fourth-largest tea-producing country according to global production statistics and the country has produced 318,470 tons of tea which contributed nearly 9.1% of the world's total tea production [11].

Nevertheless, country has dedicated over 221,000 hectares or approximately 4% of the total land area for tea cultivation. Accordingly, Sri Lankan tea plantations can be categorized as large plantations as well as smallholdings and there are approximately 118,275 hectares of smallholdings

[12] out of the total land cultivated, around 43% is managed by the corporate sector, with a production of about 35%, while the balance of 57% is in the smallholder sector, with a production of 65% of the total. The average productivity in the smallholder sector is substantially higher, at about 1,853 kg/ha compared to the corporate sector productivity of 1,459 kg/ha [9]. The production grows at an annual rate of around 10% while current average production amounted around 315 million kg with majority of black tea. The annual production contributed from various parts of the country with figures of low grown 60%, mid-grown 16% and high grown 24% respectively, while 95% of the produce were accounted as orthodox black tea [13].

Sri Lankan tea industry operates throughout the year while manufacturing tea with peak seasons in monsoons and the growing areas are mainly concentrated in the central highlands and southern inland of the country. These areas are broadly categorized as high grown tea which are ranging from 1200 m upwards; medium grown tea which are growing in elevation between 600 m to 1200 m and the low grown tea, that are grown from sea level up to 600 m altitude. The most renowned tea is produced in high grown areas which are famous for their taste and aroma. In addition, there are two types of seasonal tea produced in highland areas, which are Dimbula and Nuwara Eliya. They are much sought-after by blenders in tea-importing countries. Uva teas from the Eastern Highlands contain unique seasonal characteristics and are widely used in many quality blends, particularly in Germany and Japan. On the other hand, medium grown tea varieties are basically provide a thick liquor colour which is popular in Australia, Europe, Japan and North America and the low grown teas are mainly popular in Western Asia, Middle Eastern countries and CIS countries [9].

However, many of tea plantations have low yields because they are decades past their prime and need to be replanted [14]. The country has also shunned automation to preserve the quality of its tea. As a result, the cost of production of tea in Sri Lanka is the highest when compared with tea producing countries in the world [15]. While most producers in Kenya, Vietnam and India make black tea using a speedy manufacturing process known as cut, tear and curl (CTC), Sri Lanka favours a traditional approach that takes longer to produce more varieties [16]. Currently, Sri Lankan tea planters and manufacturers are facing intense competition from immerging countries in the global market, which has outperformed the productivity barriers and cost of production while manufacturing high quality tea in both orthodox and non-orthodox models which lead them to offer very competitive low prices to the end user [14].

The major reasons for the said drop outlined high labour costs incurred and poor food safety practices in the industry. Considering the Sri Lankan tea industry, high production costs were due to the low worker, land and factory productivity, worker shortage and escalating input costs including energy costs. The other major obstacle faced was high labour turnover [15]. There are a number of challenges which are faced by the tea industry in Sri Lanka due to current global demand and supply issues. Thus, one of the biggest challenges faced by the tea industry today is the increased cost of production, which creates number of different problems while manufacturers trying to achieve highest price from tea trading. The ever increasing energy cost coupled with fuel price hikes is another challenge. Labour shortage and labour turnover is very high, while tea growing is highly labour intensive agronomic practice which requires many labourers for day-to-day operations and tea processing also a high labour intensive product that requires many workers for various operations.

On the other hand, current labour force do not like to work in tea fields or tea factories, because of many other attractive easy jobs are available in other trades than tea with more attractive remunerations. The tea auction also has created different challenges, where tea's product quality is not a mandatory requirement to achieve a better price because buyers are not observing the factories where the product is manufactured. Nevertheless, market is itself a challenge, most of the time prices keep fluctuating and sometimes it goes below the cost of production due to the world competition. In addition, the prices of agro-inputs have skyrocketed in recent years. Most of the agrochemicals are not available or manufactured locally and are imported to the country which depends on what is happening globally.

On the contrary, counterfeit or adulterated teas are on the market which is being fought through sensitization of consumers where risks of food safety are becoming critical to the tea industry, i.e., addition of ferrous and sugar. The stifling global competition with low cost, high quality, food safety and mainly selling through the international market have being a critical issue which is ever widening the gap between other competitors in international trade. In addition there are number of certifications are accepted by different buyers which is different from one buyer to another according to the consumer market where additional costs has to be added to the cost of production, in order to strengthen the organizational reputation and to become competitive in food safety and quality aspects.

In the context of globalization, the Sri Lankan tea industry has to overcome such issues with help of the latest technology while raising the productivity in tea cultivation and processing. In addition, the recent developments in laws and legislations on food safety and hygiene practices have further led the local tea sector to a complex situation. International Standardization Organization (ISO), World Trade Organization (WTO) and guidelines of FAO/WHO Codex Alimentarius Commission (Codex) for food safety have opened up new dimensions which can be used as non-tariff barriers for the import control in consuming countries, which is becoming mandatory in many parts of developed world. The new challenge needs professionals with expertise in standards and food safety legislations to implement such requirements in the industry.

ISO 22000, FSSC 22000 are newly introduced standards which strongly focus on food safety with harmonizing all other aspects of food safety related issues and merging several food safety standards together to minimize the application of several standards for different customers. But several research outputs have revealed that, most of the tea manufacturers in the country have HACCP systems and they have been certified for ISO 22000 FSMS. In addition, there are vigorous certification systems available in most of tea factories, most instances there are no good PRP or GMP programs to be practiced because which is the foundation and one of the main prerequisites for any quality or food safety management system. Therefore, companies are compelled to go for a reliable system, however they are lacking necessary resources to materialize this task because, it is anew to the world and rather, availability of published data on tea industry related systems are also very limited. Thus, there is a big ambiguity over the credibility of the certified systems as well as there may be contradictory views over the credibility of the certification bodies in the country.

Since these standards are anew and accepted by all over the world without any resistance, it is already a driving force against exporters in international trade. However, their sister standards on traceability, and allergens were introduced into the systems which also became a stumbling block for the exporters in retaining in the international market unless they adapted to practice the current systems.

With regards to the tea industry, there is a demand for system certifications such as Japanese 5S system, ISO 9001, ISO 22000, FSSC 22000, Hazard Analysis and Critical Control point (HACCP), Good Manufacturing Practices (GMP) etc. The aforesaid two systems improve the productivity and quality whereas latter four ensure the food safety of the products manufactured, which are internationally accredited certifications based on Codex principles to ensure food safety. Accordingly, it is very important to ensure hygienic conditions in tea cultivation as well as in tea processing in order to eliminate or control hazards before they become critical to consumer health [17]. Hence, the local tea manufacturers need competent professionals to develop such systems and the tea industry has openings for consulting on food safety and hygiene applications as they do not have expert staff in these fields.

Since tea is the most commonly drunk beverage in the world with a market of 1.68 million metric tons of made tea per year [18], there are different players who dominate the multimillion dollar industry due to its economic importance, especially the west who popularize and controlled tea industry from colonial era. Considering the Sri Lankan context, the tea industry is the second largest foreign exchange earner after apparels and employs millions of workers where improvements to the tea industry would transform the country's future greatly with more sustainable development.

2
The Tea Industry

Consumption of tea began as a medicine and grew into a beverage in China, in the eighth century [19]. Tea is an evergreen bush of the genus Camellia, from south-eastern Asia, whose leaves were subject to adequate preparation with only the most aromatic, young, top two leaves and the unopened leaf bud. Up to 80,000 hand-plucked shoots were needed to produce one pound of top grade tea. The production of tea has high labour-intensive steps while processes undergo, which determine the resulting tea's characteristic colouring, taste and composition where every step has to be monitored to achieve superior quality [20]. Tea is grown not only in Asia but also in many other tropical countries, but the main exporters are China, India, Sri Lanka, Kenya and Vietnam. Consumption of tea has proven to be beneficial to health and longevity due to the compounds such as antioxidant, flavanol, flavonoids, polyphenols, and catechins content where number of developing countries benefit from growing tea as an important commodity in terms of jobs and export earnings [9].

Important Milestones in World Tea Trade
2737 B.C.E.: Tea first discovered in China by the Second Emperor, Shen Nung, known as the Divine Healer.

350 A.C.E.: The first description of drinking tea is written in a Chinese dictionary.

400 – 600: The demand for tea rose steadily. Rather than harvest leaves from wild trees, farmers began to develop ways to cultivate tea. Tea was commonly made into roasted cakes, which were then pounded into small pieces and placed in a china pot. After adding boiling water, onion, spices, ginger or orange were introduced to produce many regional variations.

479: Turkish traders bartered for tea on the Mongolian border.

618 – 906: T'ang Dynasty. Powdered Tea became the fashion of the time. Nobility made it a popular pastime. Caravans carried tea on the Silk Road, trading with India, Turkey and Russia.

780: Poet Lu Yu, wrote the first book of tea, making him a living saint, patronized by the Emperor himself. The book described methods of cultivation and preparation.

805: The Buddhist monk Saicho brought tea seeds to Japan from China.

960 – 1280: Sung Dynasty. Tea was used widely. Powdered tea had become common. Beautiful ceramic tea accessories were made during this time. Dark-blue, black and brown glazes, which contrasted with the vivid green of the stirred tea, were favored.

1101 – 1125: Emperor Hui Tsung wrote about the best ways to make whisked tea. A strong patron of the tea industry, he had tournaments in which members of the court identified different types of tea. Legend has it that he became so obsessed with tea he hardly noticed the Mongols who overthrew his empire. During his reign, teahouses built in natural settings became popular among the Chinese.

1191: Eisai Myoan, the monk who brought Zen Buddhism to Japan, returned from a trip to China with tea seeds, which he planted on the grounds of his temple near Kyoto. Eisai experimented with different ways to brew tea, finally adopting the Chinese whisked tea.

1206 – 1368: Yuan Dynasty. Genghis Khan and Kublai Khan conquered Chinese territories and established a Mongolian dynasty in power for more than a century. Tea became an ordinary drink, never regaining the high status it once enjoyed. Marco Polo was not even introduced to tea when he visited.

1211: In Japan, Eisai wrote a small book on tea, elevating its popularity further.

1368 – 1644: Ming Dynasty. People again began to enjoy tea. The new method of preparation was steeping whole leaves in water. The resulting pale liquid necessitated a lighter color ceramic than was popular in the past. The white and off white tea-ware produced became the style of the time. The first Yixing pots were made at this time.

1422 – 1502: The Japanese tea ceremony was created by a Zen priest named Murata Shuko, who had devoted his life to tea. The ceremony is called Cha No Yu, which means "hot water for tea."

1610: The Dutch brought tea to Europe from China, trading dried sage in exchange.

1618: Chinese ambassadors presented Tsar Alexis with a gift of several chests of tea.

1657: Tea was first sold in England at Garway's Coffee House in London.
1661: The Taiwanese began to drink wild tea.

1662: Charles II took Catherine Braganza of Portugal as his wife. They both drank tea, creating a fashion for it. Its popularity among the aristocracy causes alcohol beverages to fall from favor.

1669: Close to 150 pounds of tea were shipped to England.

1689: Traders with three hundred camels traveled 11,000 miles to China and back in order to supply Russia's demand. The trip took sixteen months.

1697: In Taiwan, settlers of Formosa's Nantou County cultivated the first domestic bushes. Dutch ships carried the tea to Persia, the first known export of Taiwanese tea.

1705: The yearly importation of tea to England grew to approximately 800,000 pounds.

1710: Wealthy American Colonists developed a taste for tea.

1773: The Boston Tea Party, protesting high taxes that England levied on tea, began the American Colonies' fight for independence. Under cover of night, colonists dressed as Native Americans boarded East India Company ships in Boston Harbor. They opened chests of tea and dumped their contents into the water. This was repeated in other less known instances up and down the coast.

1776: England sent the first opium to China. Opium addiction in China funded the escalating demand for tea in England. Cash trade for the drug increased until the opium wars began in 1839.

1835: The East India Company established experimental tea plantations in Assam, India.

1834: An Imperial Edict from the Chinese Emperor closed all Chinese ports to foreign vessels until the end of the First Opium War in 1842.

1838: A small amount of Indian tea sent to England was eagerly consumed due to its novelty.

1840: Clipper ships, built in America, sped-up the transportation of tea to America and Europe, livening the pace of trade. Some ships could make the trip from Hong Kong to London in ninety-five days. Races to London became commonplace; smugglers and blockade runners also benefited from the advances in sailing speeds.

1856: Tea was planted in many areas of Darjeeling.

1857: Tea plantations were started in Ceylon, though their tea would not be exported until the 1870's.

1869: A deadly fungus wiped out the coffee crop in Ceylon, shifting preference from coffee to tea.

1869: The Suez Canal opened, making the trip to China shorter and more economical by steamship.

1870: Twining of England began to blend tea for consistency.

1900: Trans-Siberian railroad made transport to

Russia cheaper and faster. Java became an important producer as well.

1904: Richard Blechynden created iced tea for the St Louis World Fair.

1908: Thomas Sullivan invented tea bags in New York, sending tea to clients in silk bags which they began to mistakenly steep without opening.

1910: Sumatra, Indonesia grows and exports tea. Soon thereafter, tea is grown in Kenya and other parts of Africa.

1970: The Taiwanese government encouraged its population to drink tea, revitalizing tea culture on the island.

Ecology of Growing Tea

Tea was believed to be originated in the highlands of south-west China, Myanmar and north-east Indian regions and the natural habitat of tea was initially the undergrowth of subtropical forests before it became commercial commodity. Modern commercial tea plantations are being cultivated longitudes between 42° N (Russia) and 27° S (Argentina), while altitudes starting from the mean sea level zero from up to 2200m. Tea plant has wide adaptability which grows in a range of different climates and soils in several parts of the world. Tea also can adapt different rain shower patterns at different annul precipitation levels with the minimum annual rainfall considered necessary for the successful cultivation of tea is 1,200 mm, while the optimum ranges between 2,500 and 3,000 mm. Rainfall must have to be evenly distributed year-round to get an optimum yield with an annual average temperature around 18-20°C which is usually ideal for the tea bush. In addition, the soil type must be deep, well drained and exhaustively aerated, nutritious red-yellow soil with a low pH (4.5-6.5) where extended drought periods, water logging conditions and temperatures below 12°C and above 30°C are not favourable for the growth of tea [21]. The slope must not be too steep and the maximum tolerable gradient is 25 degrees.

Botany of Commercial Tea Plant

The locally grown tea plant, which was botanically known as *Camellia sinensis (L.) O. Kuntze*, was belong to the plant family *Theaceae*. Today cultivated varieties were hybrids of the original tea plants *Thea sinensis* and *Thea assamica*; for botanical reference the two extreme varieties *Camellia (Thea) sinensis var. sinensis* and *Camellia* (Thea) *sinensis var. assamica* are distinguished. The variety *sinensis*, also called China tea, is suitable for growing in marginal areas of the subtropics. It is more droughts tolerant and can survive short frost periods. The leaves are small (up to 9 cm long), tough, leathern, with edges, not pointing, with a strong flavour and less yielding. The variety *assamica*, also called Assam tea, is a tropical variety, sensitive against dryness and cold weather conditions. The leaves are big (up to 35cm long), soft, without edges, pointing, with less flavour and high yielding. Under natural conditions tea grows as an evergreen tree 8 – 15m tall with a strong taproot growing about 6m deep [22].

Tea plants are propagated in two different ways which can be reproductive by seeds or vegetative by young tea branch cuttings. Vegetatively propagated tea plants are referred to as VP and seed base plants are called as seedlings. Seedling tea plants only can develop an extensive taproot which is helpful in surviving drought condition where seedling tea is said to be low yielding, hardy and does not give a uniform stand of yielding or leaf types because of the heterogeneous source planting material. On the other hand, the VP plants are more standardized for the uniformity of raw materials. Nevertheless, the standardization of vegetative propagation (VP) has offered an economic method to provide uniform quality planting material with a large choice of improved clonal tea varieties of high yield potential. Tea clones are appeared to be analogous in their gross morphology to one another because of their origin from closely related species of Camellia cultivars. However, the growth and development of the tea plants in field conditions are influenced by eco-climatic system, inherent vigour and cultural operations. Thus, different clones of tea species exhibit variations due to their inherent qualities, besides nutritional and hormonal factors even in the nursery conditions [23].

Nevertheless, VP tea cultivars are proven to offer its typical characteristics while giving a uniform cultivation, with common disadvantages of more sensitive to the climatic changes, nutrient supply, weeding, pruning and demanding more intense conditions regarding the cultivation than seedling type tea cultivars. In general, Seedlings or VP cuttings are raised under controlled conditions of shade and humidity over a period of 1-3 years in a nursery. The young tea plants were initially trained to become a bush with a cut across at about 23cm above the ground to encourage lateral branch growth after tea cultivars are planted in the field which is repeated at different growth stages to provide formation of a bush. The tea plants are considered to be immature until the first light plucking of young shoots takes place; depending on the plants' condition. The plucking is normally started in the fourth year where it started to refer to as mature tea plant.

Varieties of Made Tea
All truly fine teas have in common that only the most aromatic, young, top two leaves and the unopened leaf bud are used. Up to 80,000 hand-plucked shoots are needed to produce one pound of top-grade tea. The production of tea is a labor-

intensive process and every step is essential to achieve superior quality.

The tea that we drink every day, though it occurs in many varieties, is made from the leaves of the same tea bush. The secret of species variety is actually in the manner of processing collected leaves. The processes they undergo, determine the resulting tea's characteristic colouring, taste and composition.

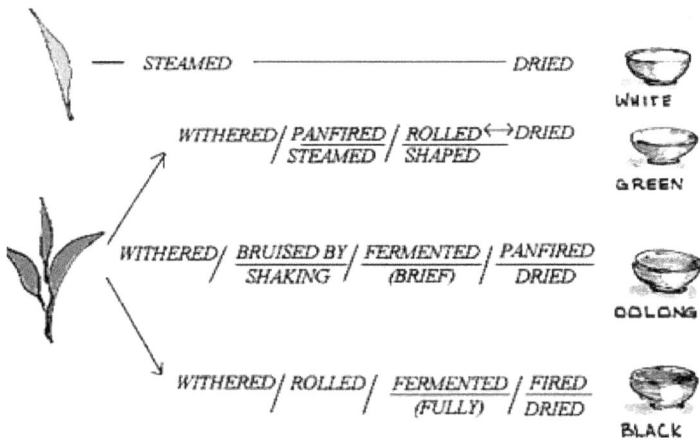

Figure 2.1 Classification of tea varieties

Black Tea

Currently, the most popular kind of tea in Europe. It is made as a result of withering, rolling, fermentation and drying (precisely in this sequence). According to a legend, it was created accidentally, when during transport to England by ship, a load of green tea was soaked, which resulted in its fermentation (oxidation).

The most popular varieties of black tea are:

Yunnan – originating from China, it owes its name to the province where it is cultivated and which is considered the cradle of tea

Assam – traditional Indian tea with strong taste and characteristic copper coloured infusion

Darjeeling – Indian tea with an exceptionally mild taste. It is often called the champagne of teas. It is best on its own with nothing added (e.g. milk) that could spoil this exceptional tea

Ceylon – tea produced in Sri Lanka, it has a clear taste and red and brown colour of infusion. The major variety is orthodox black tea which represents the 33% of world requirement. Other varieties are CTC and rotorvane orthodox black tea. Considering the "Pure Ceylon" as a world renowned brand, there are several distinctive black tea varieties are exported. The following classification is not comprehensive, but it provides some of the major grades and products of the country.

Black Tea – Main Grades
Pekoe is a whole leaf black tea grade produced by a medium plucking of the second leaf on the tea bush where the term Pekoe is derived from Chinese language, meaning 'white hair' and was originally

applied to early tea pluckings, due to the white down on the backs of the young tea leaf.

P/PEK – Pekoe: Twisted and coarse.

Pekoe1 –Pekoe one: Twisted and coarse, but small in size than the Pekoe.

OPA – Orange Pekoe 'A': bold, long leaf tea which ranges from tightly wound to almost open which is a good quality tea, consisting of large and slightly open leaf pieces.

OP – Orange Pekoe: Same style but small than OPA and it is in-between OP1 and OPA, which is a main grade, can consist of long wiry leaf without tips.

OP1 – Orange Pekoe One: slightly delicate, long, wiry leaf with the light liquor and more wiry than OP.

BOP – Broken Orange Pekoe: Black tea comprising smaller leaves and broken pieces with an abundance of tips. BOP is identical for the strength and tannic acid taste.

BOP1 – Broken Orange Pekoe one: Wiry and small than OP1.

BP – Broken Pekoe: Full boiled black tea comprising broken segments of somewhat coarse leaf, without tip.

F/Fannings – Fannings are crushed leaves of less than one and a half millimeters in length which produce a strong beverage with plenty of body: Fannings will produce liquor that is often as good as that of a whole leaf grade.

BOPF – Broken Orange Pekoe Fanning's: Smaller than BOP leaves, broken leaf, slightly larger than dust.

FBOP – Flowery Broken Orange Pekoe: Coarser and broken leaves which is similar to BOP, but slightly bigger in size and consisting few tips.

FBOP1 – Flowery Broken Orange Pekoe one: Little bit smaller than the BOP1.

FBOPF – Flowery Broken Orange Pekoe Fanning's: Similar to the BOP leaf but firm leaf and consist few tips.

FBOPF1 – Flowery Broken Orange Pekoe Fanning's one: Similar to the BOPF but firm leaf and consisting little more tips then FBOPF.

FBOPFSP – Flowery Broken Orange Pekoe Fanning's Special: Similar to the FBOPF1 but firm and more black leaf with much better tips, prices are high.

FBOPFEXSP – Flowery Broken Orange Pekoe Fanning's Extra Special: Similar to FBOP1 but firm and more black leaf with much better leafy tips, expensive variety.

D1 – Dust one: The smallest of particles smaller than Fanning's leaves.

D – Dust: Similar to D1 but will appear slightly brown powder.

D2 – Similar to D but will appear fibrous brown powder leaves price must be low then D.

Off Grades - Teas which are is not match for main grades.

BOP 1A – Broken Orange Pekoe one A– Weight Less Large leaves.

BM – Broken Mix: Smaller then BOP 1A's.

FNGS – Fanning's broken leaves, which is slightly larger than dust.

FNGS 1 – Fanning's broken leaves, which is slightly larger than FNGS.

PFNGS – Fanning's broken leaves, which is slightly larger than dust.

PFNGS 1 – Fanning's broken leaves, which is slightly larger than PFNGS.

BP – Broken Pekoe: Broken stems.

BT – Broken Tea: Broken pieces.

CTC Teas – crush, tear, curl or cut, tear, curl
CTC tea consists of leaves that have been cut, torn and curled rather than rolled as in the conventional orthodox method of tea production, which is done by machines and it speeds fermentation and reduces production time. CTC tea is mainly use for tea bags.

BPS – Broken Pekoe Special: Even curl pieces

BP1 – Broken Pekoe one: Little smaller then BPS.

BPL – Broken Pekoe Leaf: Even leaf pieces.

PF1 – Pekoe Fanning's one: Similar to BP1 but small pieces.

PD – Pekoe Dust: The smallest of particles smaller than PF1 granules.

Garden Teas

There are several garden teas, which can be further categorized as single origin teas in most of the cases. There are several tea estates and factories have won the international consumer's preference due to distinctive taste and aroma generated during seasonal weather fluctuations as well as manufacturing processes.

Lover's Leap FBOP – Lover's leap estate is one the best tea gardens which produce FBOP with delightful flavour that can be explains as astringent and lively during the peak season.

St. James FBOP – The FBOP produced in St. James estate has distinctive flavour and pungency. The plantation is located in Badulla district where extreme weather conditions are generated due to the topography of the landscape and rain pattern which creates its distinctive flavours.

Inverness OP – This OP are from Nuwara Eliya which are grown in mountain ranges above the clouds and are often refers as Champaign of teas.

Pettiagalla OP – The Pettiagalla estate is famous for this variety which is high grown tea and that are broken leaves with elegant, long and wiry particles. The brew is golden yellow with flowery flavour which is one of the most aromatic teas in the country.

Uva Highlands FBOP – Uva highland teas can be distinguished from other tea varieties because they have a significant characteristic of being full-bodied with a spicy and bit of a sparkling flavour.

Kenilworth OP1 – The Kenilworth estate is another highland tea plantation which produces long, wiry tea with slightly sweet flavour and amber coloured excellent body in the brew.

Earl Gray – The Earl Gray is a blend where black tea is infused with bergamot orange to get a distinctive flavour and aroma which has flowery flavour and aroma.

Dimbulla BOP – Dimbulla region is famous for strength and dark liquor colour which is slightly sweeter in taste.

Mahagastota PEKOE – The Mahagastota PEKOE has a light yellow liquor colour with superbly balanced high grown flavour and aroma due to young tea leaves and buds.

Green Tea
In contrast to black tea, during the production of green tea the fermentation stage is skipped, and its leaves are exposed to steam for 30-45 seconds. As a result of which, they retain more beneficial substances, such as vitamin C and minerals. In order

not to destroy these substances, green tea should be infused with water at the temperature of 85-90°C, and not with boiling water (100°C) as you would do with black tea. The drying of green tea at a very high temperatures (baking) gives it a specific taste and aroma, and also enables longer storage.

The most popular varieties of green tea are:

Genmaicha – made in Japan, its characteristic taste is the result of mixing proper tea with grains of brown, roasted rice.

Gunpowder – Chinese green tea with characteristic leaves rolled into balls, to which it owes its name. During infusion, leaves unroll, giving a greenish, coppery infusion with a slightly bitter taste.

Sencha – Japanese tea with long, relatively thick leaves. It contains a large dose of vitamin C.

Gyokuro - Japanese dew drop, leaves in the form of pointy, emerald needles that give a yellow green infusion. Bushes before harvest are almost completely exposed to sunlight that is why they are dark green with a large amount of chlorophyll and little tannin.

White Tea
White tea is obtained exclusively from leaf buds. Its homeland is plantations in the Chinese province of Fu-Tien. The production of white tea is tightly

controlled and is conducted according to a rigorously observed, traditional method. The withering and drying of white tea is fully natural, which affects its somewhat different appearance, taste and bouquet. Its Infusion is noticeably lighter than with green teas. White tea also contains the most theine compared to all other varieties, which is why it has such strong refreshing and invigorating properties. Small admixtures of white tea clearly improve the taste and bouquet of black teas.

Pai Mu Tan - from special Big White tea bushes only buds are collected.

Red Tea (Oolong, puszong, Pu-erh)
Because of the way it is produced, red tea is sometimes called 'under-fermented' or 'semi-fermented'. After picking, leaves are dried and then put into bamboo baskets and shaken in order to graze the edges. Then the leaves are fermented, but fermentation is shorter than for black tea, thanks to which the tea retains more beneficial compounds. Depending on the duration of fermentation, different kinds of red tea are obtained: pu-erh, oolong etc.

Herbata Mate (tzw. Yerba Mate)
Mate tea is a unique variety, originating from South America, and more precisely from Argentina. Its Infusion is made from the leaves of Yerba mate. It contains, similarly to green tea, many valuable

vitamins and minerals. Yerba Mate improves physical and intellectual performance, concentration and clarifies the mind. In addition, it helps to regulate metabolism and increases resistance to disease. It is recommended during dieting, because thanks to its nutrient content it eliminates hunger. Special vessels, made of bottle gourd or hardwood, are used in the preparation and drinking of Yerba Mate.

Yellow Tea

Yellow tea is prepared from the youngest shoots and buds of leaves that are partly (not more than 12-15%) fermented during the process of rolling. It is similar to green tea of the highest quality. In China it was called 'Emperor's tea', because its use was restricted to the Emperor's court and some religious ceremonies. In Europe, it is almost unknown and almost unavailable in trade. Principles of infusion are the same as for Chinese green teas.

Flavoured Teas

A tea can be called flavoured when it contains *Camellia sinensis* or *assamica* leaves which has had dried fruits, spices or herbs, petals or fruit essences/aromas are added during blending, all kinds of tea, although green tea especially, are used to make flavoured tea. Some teas have natural aroma, because they grow e.g. among wild orchids, others take on the smell of blossoming fruit trees. But in most cases, different kinds of fruit aromas or powdered juices are added to teas. Traditional non-tea ingredients are, jasmine flowers, rose, cornflower or mallow and pieces of dried fruits, such as mango, pineapple, orange, cherry, blackcurrant, apple or strawberry. The taste of teas can be improved by adding citrus peel (e.g. lemon). Another popular ingredient is lemon grass. Spices such as ginger, vanilla, cinnamon or ginseng, and herbs such as chamomile and mint also make a perfect complement to tea.

The most famous flavoured tea is Earl Grey which contains bergamot oil. Bergamot is a citrus, a relative of wild orange. It gives it a subtle citrus flavour which is the trademark of Earl Grey tea.

3
Orthodox Black Tea Manufacturing

Tea is a product with different colour, taste, smell as well as different shapes in visual appearance based on its type or variety, but its processing methods are almost similar with minor variations. However, the quality and the characteristics of the final product basically depend on the contents of the leaf where large number of chemical constituents is present in fresh leaf. The most important group of chemical compounds present in the leaf is polyphenols which undergo series of modifications during black tea manufacturing process to impart the specific colour and flavour of tea brew. In addition, caffeine, amino acids and carbohydrates are also include in the tea leaf, but all these chemical compounds are stored separately in different specific cells or within the leaf without mix-up to prevent chemical reactions [24] such as generation of dimeric or polymeric phenolic compounds by polyphenols in the presence of polyphenolic oxidase and oxygen. Tea will deteriorate in no time by fungus attacks, if the moisture and temperature is not carefully controlled during the manufacturing process as well as delivery to the end consumer.

Tea Plucking
First process step of the tea manufacturing is tea plucking, where tea leaves and flushes including

terminal bud with two top young leaves are picked from tea plantations and it was transported to manufacturing facilities in ventilated trucks under loosely packed conditions [25]. The younger tea leaves contain higher concentrations of relevant chemical constituents which is ideal for the production while coarse leaves contain lower amounts where it is very important to have tender two leaves and the bud for production to manufacture good quality black tea [24]. The plucked tea leaves are subjected to leaf count at the receiving where B – 60 or "Randhalu" method is applied and if the receiving green leaves had more than 75 – 80% of tender shoots with two young leaves and bud will ensure the better quality of the final product [26]. The plucking is manual and it was carried out by hand which is a labour intensive process where picking is performed by pulling the flush with a snap of forearm, arm or even by using shoulders while gasping the tea shoot using the thumb and forefinger with middle finger or combination [24].

Withering
The first stage of black tea manufacturing is withering, which refers to the changes in green tea leaf from the time it is detached from the plant to the time of maceration [27], while chemical withering involves biochemical changes, which solely depend on time [28]. Tea leaves are subjected to weighing and scrutinizetion for quality and moisture on the

arrival and taken into withering section of the factory which is normally built on top floors of the tea manufacturing facilities. The leaves were loosely stacked in withering troughs for controlled withering with free flow of air as well as electric fans which can provide heated air flows under controlled conditions where frequent turning and mixing with supervision are mandatory. Normally, withering requires 16 to 18 hours depending on weather condition and the moisture content of the tea leaves [29].

Tea leaves are subjected to wilting just after picking from the plant with gradual onset of oxidation where withering is applied to remove excess water from the tea leaves while minimizing the enzymatic oxidation within the leaf [30] which was carried out on the withering troughs in orthodox manufacturing model and it was considered as physical wither. On the other hand, minimum of six hours is required for the desirable chemical changes [29] which have to be minimized or delayed till the leaf is ready for the rest of the processing with exposure to oxygen. The leaves are subjected to reduce more than quarter of its weight in certain processes which is very important to catalyze the breakdown of leaf proteins into free amino acids while increasing the amount of freely available caffeine, that change the flavour of tea [31]. This process was basically considered as chemical wither which starts with the breakdown of larger molecules in to smaller molecules while result in increase of amino acids, flavour compounds, caffeine as well as increase in cell wall permeability for rest of the chemical withering [24]. During this process, approximately 63% moisture is extracted from the leaves, making them soft and pliable for further processing [32]. Thus it was necessary for the operator to have control of all those important chemical and physical changes, in order to have effective control of quality of the final product [29].

Disruption/Rolling

Most of the Sri Lankan tea factories manufacture orthodox tea because country is renowned for these teas while both producer and the buyer prefer to continue orthodox character of Ceylon teas which was basically carryout using orthodox or rotorvane orthodox rollers [29]. The disruption which is also called as leaf maceration by westerners is carryout to bruise or torn the tea leaves for the promotion of quick oxidation [30] which is very important in tea manufacturing process and it was carried out using manual means or mechanical means based on the requirements. The light bruising is achieved manually while mechanical methods such as kneading, rolling, tearing and crushing are used for more extensive disruption [33] which breakdown the structures inside and outside of the leaf cells allowing to blend oxidative enzymes with various substrates to begin oxidation [34]. The leaf maceration releases some of the leaf juices inside the cells which may aid in oxidation while changing the taste profile of the tea. The rolling process is used to form damp tea leaves into wrinkled strips where tea leaf is wrapped around itself. On the other hand, rolling action further causes some of sap comes out with essential oils and juices inside the leaves to ooze out where taste of the tea is further enhanced [30]. In orthodox method, various types of rollers are used to macerate leaf where first roll is mostly used as preconditioning roll with very gentle pressure which promotes distribution of leaf juice on to the surface of twisted particles and it dry up on the surface to enhance blackness of final product. The subsequent rolling program is planned to achieve through breakdown of leaf cells [24].

The orthodox rolling process is complex and the leaves are rolled by applying mechanical pressure to break up the cells and extract the cell sap at the orthodox roller with approximately around 20 minutes, the macerated leaves; still damp from the sap are sieved on roll breakers to separate the finer leaves which is called the first dhool. The rolled tea leaf is sieved due to further breaking up of twisted leaves will disappear its twisted-up appearance. Thus prompt separation of fine twisted particles from larger sized bulk is required while controlling of the initiated chemical reactions due to cell breakage is a necessity where roll breaking is applied to effectively separate them. The extracted uniformly fine particles are called dhools [29]. These are spread out or staked on racks immediately for fermentation, while the remaining coarse leaves are rolled for a further 20 minutes under higher pressure. The rolling and roll breaking process is continued up to 3rd dhool or 4th dhool and rest of the coarse leaves which are not pass through the last sieves called the big bulk. As a matter of fact, short rolling time produces larger leaf grades, while longer rolling produce smaller leaf grades. During and after the rolling process, the cell sap runs out and reacts with oxygen, thus triggering the fermentation process while the essential oils responsible for the aroma are released [32].

Considering the orthodox black tea fermentation, during the rolling process, cell breakages causes the mixing of enzymes with other chemical compounds while polymerization primarily starts and continues through rolling and roll braking processes. Further, the fermentation is allowed to polymerize on racks or on floor such time as desirable. The most important polymerization reaction that is responsible for liquor character is the oxidation of polyphenols by the enzyme polyphenol oxidase and this reaction is highly temperature dependent [29]. The optimum temperature range for this reaction is 25 - 30°C [35].

Oxidation/Fermentation
The oxidation of tea represents a series of complex chemical reactions which begins just after the cell

maceration in orthodox roller where mixing up of enzymes with other chemical compounds within the cell generates number of reactions [24]. Certain teas require more oxidation while some of them need less oxidation due to the specific production methods, however orthodox black tea require 100% oxidation to convert phenolic compounds in to polyphenolic compounds. The oxidation process in tea manufacturing is also referred as fermentation where chlorophyll pigments in the tea leaves are enzymatically broken down while releasing the tannins or transforming into other compounds. The amount of fermentation required for a given tea sample is decided by the process owner based on the finish product quality required and the weather conditions because oxidation is the most important factor in the development of aroma compounds, which gives the resulting tea's liquor colour, strength and briskness [31].

Due to the maceration of tea leaves cells and the subcellular compartments disrupted where cytoplasmic polyphenol oxidase (PPO) allows oxidizing the flavan-3-ols in the vacuoles. Thus, majority of the monomeric flavan-3-ols are oxidized and polymerized to form thearubigins (TRs) and theaflavins (TFs) [36], [37]. Nevertheless, first step of the process is oxidation of polyphenol in the presence of polyphenolic oxidase where polyphenol is converted to very unstable oxidized polyphenol which further react with polyphenols to create more stable polyphenolic compounds.

There are two type of polymers produced due to these reactions; the orange red compounds are called Theaflavins (TF) which is generated by dimerization while dark brown compounds or the Thearubigins (TR) are generated due to polymerization of three or more compounds. The percentage formation of TR and TF is related to the fermentation time and the temperature of fermenting dhool where high temperatures will lead to rapid formation of TR groups rather than TF groups which will impart a dull colour to the liquor. Thus, control of temperature is of a paramount importance. There are other reactions such as oxidation of carotene and amino acids with oxidized polyphenols which contribute to the flavour of the made tea [24]. Thus, optimum temperature range for the reaction is 25 – 30°C whereas the completeness of fermentation reaction is best judged by the characteristics of final product [29] which intern find the appropriate price in the tea auction.

Firing
The next step of orthodox tea processing was drying which is carried out to terminate biological reactions by heat denaturation of enzymes while reducing the moisture content to increase the shelf life of orthodox black tea and to enhance the chemical reactions that are responsible for the character and

flavour of orthodox black tea [38]. On the other hand, firing further influence balancing of flavour of the tea, because firing eliminates some of the less desirable low boiling point compound such as volatile constituents are removed while retaining more useful high boiling point compounds [24]. The polyphenolic oxidase enzyme will convert catechins into theaflavins till it reaches the temperature 55°C where enzymatic reactions will not be arrested as soon as the fermented tea reached the dryers. Actually fermentation is accelerated at the beginning of the drying where 10 to 15 percent of theaflavin content in the black tea is formed during first 10 minutes of drying [38].

In addition, drying process reduces the most of the moisture percentage in the wet dhools which comes out of the dryer mouth reaching after 3% [29]. Nevertheless, the colour of the wet dhools changed to black from green due to the transformation of chlorophyll into pheophytin, which imparts the desired black colour of the orthodox black tea while reducing the stringency of black tea due to combination of polyphenols with tea leaf proteins at drying because of elevated temperatures. On the other hand, volatile compounds are reduced due to drying; however, some of the aromatic compounds are formed. The fermented dhools are dried through hot air circulation where it needs around 20 – 30 minute based on the specific manufacturer in conventional drying units while fluid bed dryers are operated around 20 minute with 125°C [38].

Grading

The conventionally dried teas are cooled before issued for grading while fluidized bed dryers release tea after cooling which can be directly used for sifting operations.

The sifting is carried out by sorting the leaf particles into different sizes defending on the market demand as well as buyer requirements according to their popular blended brands. The primary objective of the sorting is to enhance the value while imparting the quality. The process of sorting enhances the appearance of the liquor quality while removing the fiber or flakes of coarse leaf particles. Thus sorting is carried out in four stages which are cleaning of fiber, grading, winnowing and colour sorting [29].

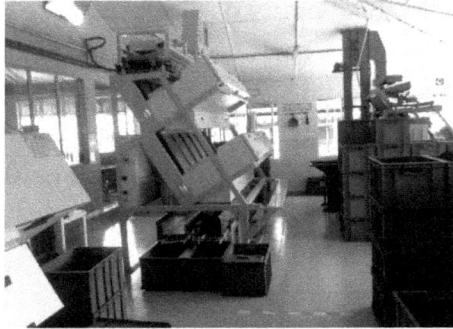

Nevertheless, orthodox black tea generally has four scales for quality where whole leaf tea is considered highest quality followed by broken leaves, fannings and dusts. Hence, whole leaf tea is produced without alteration to the tea leaf or with very little alterations where it will end up in finish product with coarser texture while bagged tea has more smooth appearance. However, whole leaf tea is as the most valuable, especially with high content leaf tips. On the contrary, broken leaves are considered as medium grade tea which is basically sold as loose tea while smaller broken grades may be used for tea bags. The next grade fannings are considered as leftover products of larger tea grades which is smaller in size and it is also suitable for the tea bags. Dust is the finest particles of leftover products of all other three types which are also good for tea bags that can be utilized for very fast harsh brews, as the greater surface area of the many particles allows for a complete diffusion of the tea into the water. Fannings and dusts have a darker colour, stronger flavor when brewed but lack sweetness [39].

Bulking
The bulking of made tea is basically carryout to even the latter dhool particles as well as early dhool particles which is very important to eliminate day-to-day variations in the produce and to increase the quantity of a single grade. Most of the orthodox manufacturers use manual methods for the small quantities manufactured due to large array of grades [29].

Packing
Tea is packed in paper sacks in current manufacturing facilities, however it was earlier

packed in plywood box few decades ago and it was the most popular method at that time. Tea may be consumed after many months of preparation where it needs preservation techniques to improve the keeping quality while preserving its desirable characters which will be deteriorate due to the absorption of moisture where packing needs special attention to resistance against moisture absorption [29].

Tea Brewing

Tea has to be brewed to get the liquor out; about one teaspoon or 2.25 grams of orthodox black tea is used per 180ml of water in 6 ounce cup where it must be steeped in freshly boiled water unlike green teas which turn bitter when higher temperatures were used. On the other hand, whole leaf black teas need to be steeped around 4 to 5 minutes, but broken leaf grades require less brewing times because they have more surface area than whole leaf grades. In addition, Darjeeling tea requires 3 to 4 minutes of steeping to have a better brew [39].

4
Story of Ceylon Tea

The commercial planting of tea in Sri Lanka was introduced by a Scotsman, James Taylor in 1867 [8]. Since then the Ceylon Tea has been the world's number one brand for over a century and the local tea industry has made a significant growth over the time while securing its position in the global market as a leading producer and exporter of high quality black tea. In terms of international trade, tea is one of the major export revenue earners for the country, where thousands of lives are depending on it directly or indirectly. In 2007, Sri Lanka was the fourth-largest tea-producing country according to global production statistics and the country has produced 318,470 tons of tea which contributed nearly 9.1% of the world's total tea production [11].

Considering current global tea production statistics, CTC tea has the lead with 44% of the total global production while orthodox black tea production was 31% out of total and rest was covered by the green tea. Since Sri Lanka manufactured approximately 95% orthodox black tea which basically intended for export representing 32% of the global demand on orthodox black tea [41].

According to the Sri Lankan tea history, 205 tea plants of Camellia sinensis var. assamica were brought from India to the Peradeniya Botanical Garden in 1839. The Chinese variety of tea plants were planted in Pusselawa two years later and commercial tea planting was began in 1867 [42] after the rust leaf disease due to *Hemileia vastatrix* which destroyed Sri Lankan coffee plantations [43]. The scotch man James Taylor was became the first commercial tea planter in Sri Lanka, who planted approximately 8 ha at Loolecondera in Hewaheta in Kandy district in 1867, which is still in production [42, 44].

There was a massive labour requirement for clearing forest vegetation for commercial cultivation of tea at the time, which was achieved through bringing Tamil labourers to the country from south India (Tamil Nadu). The first consignment of "Ceylon Tea" was reached to London in 1873 under the British colonial time, and English government setup necessary infrastructure while introducing plantation management to the country. But British Empire was ceased in 1948 after the World War II due the changes taken place in British political stream as well as the national freedom struggle. The country was freed from the British Empire and national socialist movement became stronger where government officially commenced nationalization of plantations in 1958 and most of the foreign investors left the country seeking plantation investments in other countries like Kenya [21]. In 1972, the leftism and socialism became very stronger within the political system and all the private tea plantations which were mainly managed by the British agency houses and few local individuals were ruled out of the industry with the nationalization of entire tea plantation sector except small scale planters who manage less than 50 acre land slots.

The mismanagement of nationalized plantations and corruption was identical in the sector while government accumulating the huge losses. On the other hand, small percentage of nationalized lands was distributed among the Sinhalese population after the land reformation act in 1972/75 for housing and agricultural purposes as a part of the objectives of land reforms act which was to distribute lands to the people who don't own lands. Other objectives of the same act were to solve unemployment problems as well as to increase the agricultural output of the country while cutting down imports which popularize the government due to the free distribution of lands where over 325,000 persons or 20% of the people employed in subsistence or small scale agriculture in villages were benefited from the program by 1975 [45].

The smallholder farmer has been defined as "cultivated tea land of less than 10 acres (4.05 ha) which has the plant density higher than 1000 per acre under a possession of an independent individual cultivator was called as smallholder while his tea garden was defined as smallholding and if the cultivated tea land was larger than 4ha has been called as an estate [46]. On the other hand, if the considered land has less than 1000 tea bushes per acre, it has been considered as abandoned. Most of the gifted lands were in the size of 0.2 – 0.4 hectares which later promoted smallholder famer population in the country because people use the given land plots to build a house as well as a small home garden for household vegetable supply before they utilized it for commercial cultivations. The lands became economically insufficient for tea cultivation due to that and this was further enhanced due to the marginal lands with steep slopes as well as unavailability of necessary infrastructure for new farmers such as extension, marketing, credit system, etc., where land reformation act was primarily focused on the forth coming elections [47].

Considering the Sri Lankan tea industry, country's geographic location and the topography was dynamic where climatic and agro-ecological conditions were based on the same factors which indirectly define the tea growing areas. Accordingly, entire wet zone and some considerable parts of the intermediate zone such as Uva basin and eastern parts of the Badulla district are compatible with requirements of sustainable cultivation of tea [45]. The growing areas were distributed differently and mainly concentrated in the central highlands and southern inland areas of the country. Thus, tea fields or growing areas can be classified in to 3 major categories which were also used to sell tea in the auction as well as in the global tea trade. The classification based on the elevation of the growing area where above 1200 meters from mean sea level is considered as high grown or up country tea while 600 meters to 1200 meters are considered as medium grown or mid country tea. The low grown or low country was below the 600 meter down to the mean sea level [21].

The high grown tea of the country has high reputation around the world due to the taste and aroma. In addition, major characteristic differences in tea quality and yield as well as seasonality are directly proportional to the elevation and climatic factors which directly define the tea characteristics such as flavour, liquor and colour, when tea grades were offered for sale and prices fetched in the auction. The specific areas were Dimbula, Nuwara Eliya and Uva highlands where seasonal changes in made tea was identical and highly sought after by different blenders around the world for high quality blends in international markets [29] such as European community and Japan. On the other hand, medium grown teas provided thick liquor colour which has created different specific market segments such as Australia, Europe, Japan and North America while low grown leafy grades with well twisted appearance were basically popular in Western Asia, Middle Eastern countries and

Commonwealth of Independent States (CIS) countries [41].

Sri Lankan Tea Statistics

Sri Lankan tea industry annually produced around 320 million kilograms of made tea according to the current statistics available. Out of the given production output, country has manufactured approximately 95% orthodox black tea annually which basically intended for export representing 32% of the global demand on orthodox black tea [48]. The CTC and green tea represents only 5% of the total production and country manufacture tea throughout the year.

Nevertheless, country has dedicated over 221,000 hectares or approximately 4% of the total land area for tea cultivation. Accordingly, Sri Lankan tea plantations can be categorized as large plantations as well as smallholdings and there are approximately 118,275 hectares of smallholdings [12] out of the total land cultivated; around 43% is managed by the corporate sector, with a production of about 35%, while the balance of 57% is in the smallholder sector, with a production of 65% of the total. The average productivity in the smallholder sector is substantially higher, at about 1,853 kg/ha compared to the corporate sector productivity of 1,459 kg/ha [9]. The production grows at an annual rate of around 10% while current average production amounted around 315 million kilograms with majority of black tea. The annual production contributed from various parts of the country with figures of low grown 60%, mid-grown 16% and high grown 24% respectively, while 95% of the produce were accounted as orthodox black tea [13].

According to the Sri Lanka Tea Board, tea smallholder's concentration was 21.6% of the total land area cultivated in Galle, 20.8% in Matara, 20.5% in Ratnapura and 12.4% in Kandy where tea smallholder share on the cultivated tea lands in low elevations were higher than the total tea lands cultivated by the estate sector [46]. Further to that, in 2005 Department of Census and Statistics published the census on tea smallholdings showing that the contribution from the corporate sector was declining from 1990s while smallholder sector contribution was increasing where low-country and mid-country production trends were increasing with the number of smallholdings amounted to 397,233 and around 80% of the holdings were from low country. The number of smallholders was amounted to 370,842 with land extent of 118,275 hectares which produced 204 million kilograms per year with 88% of highly productive vegetatively propagated tea plants and the rest (12%) with seedling varieties [12]. Nevertheless, the corporate sector owned 78,188 hectares of cultivated tea lands with 47% of highly productive vegetatively propagated tea plants and rest (53%) of the plantations were old seedling types which has contributed 114 million kilograms to the national production [49]. Considering the green leaf supply to finish product manufactured; 1 kg of made tea was prepared using 4.50 to 4.65kg of green leaf [41].

According to the statistics, country's annual production rate was growing at around 10 million kilograms a year over a last decade with favourable weather conditions, use of vegetatively propagated varieties for planting materials, expansion of smallholder community, privatization of estates and use of proper management practices has been identified as the factors affected for the growth of the annual production rates. In 2005 country produced 316 million kilograms of made tea which can be segregated according to the elevation such as 181 million kilograms in Low country (0 – 600 m above mean sea level – amsl), 55 million kilogram of mid country (600 – 1200 m amsl) and 80 million kilograms in up country (> 1200 m amsl) [41].

Problems Encountered in Tea Industry

The current production is having number of problems where field productivity is very low which has direct impact on the productivity of the manufacturing process as well as labour. The field productivity was low basically due to the eroded soils with lower nutritional quality because slope of the land is higher and most of the tea lands are located on steep areas where top soil is completely eroded.

On the other hand, tea bush per area is lower in seedling type tea plantations where productivity is further low while plant needs more space. Nevertheless, cost of production is very high compared to the other tea producing countries due to the increased worker wages, fuel prices, electricity costs, firewood prices as well as increase in brought leaf competition which has created a situation where cost of production has far beyond other competitive nations. The trade issues in productivity has further intensified due to yield decline and bush debilitation in certain parts of the tea growing regions (eg: Deniyaya and Uda pusselllawa) as well as incidence of dry weather pests during wet weather which is probably due to the global warming, affecting the productivity of tea plantations [21].

Labour Issues
Considering Sri Lankan tea industry, it lacks qualified staff to supervise, both cultivation and processing where, the caliber of factory and field staff in service too needs to be strengthened. Nevertheless, Planters Association of Ceylon has recently revealed that, there is exodus of especially Managers, from the sector. The industry further compressed due to scarcity of skilled workers which is similar to the shortage of staff, the industry has less skilled workers for several cultivation practices such as pruning, weeding and fertilizing and also to

perform certain specialized duties in the factory, such as operation of the critical equipment. There is severe worker shortage in both sectors, i.e., the corporate sector as well as the smallholder sector, where the problem is so acute, thus industry has already compelled to mechanize its field and factory operations. Hence fears are expressed, whether Sri Lanka could still maintain its image as the producer of best quality tea.

The Cost of Production
Sri Lanka hold the fourth position in global tea trade as a leading manufacturer with the 9% stake of international market while export demand for the country's tea was around 19% on the global scale. The country's major problem to the improvement in production and infrastructure was cost of production (COP) which was highest among tea producing countries where profitability is less compared to the other tea producing counterparts in the world [48]. The high cost of production can be impacted negatively on the industry because that can create economic turmoil which can hit the poor people's stomach while cutting down exports. Nonetheless, such kind of situations were very critical due to the lower profit margins entertained by the manufacturers who intern adulterate or letdown the quality and the good hygienic practices while providing less benefits to the workers where everything collapsed to breakdown the branding Ceylon Tea has earned over the years in long run.

The corporate sector has highest COP which basically depend on the cost of green leaves and the cost of green leaf again depend on the productivity of the field, wages, plucker intake, cost of other inputs such as weeding, fertilizing and transportation costs etc. The current rate of a green leaf 1kg was around LKR 75.00 – 80.00 and plucking cost was around LKR 25.00. Thus total gross cost of the green leaf 1kg was around LKR 60.00 where corporate sector has lower profitability rates due to the many other overheads accumulated while manufacturing process up to the auction. On the other hand, smallholder sector operates in different models where household labour mostly utilized in many of the crop management and plucking operations as well as transport.

Accordingly, their profitability ratios were somewhat higher with high productivity rates and most of these leaves were processed in private brought leaf factories around where the Tea Control Act has been stipulated for the reasonable price for purchased green leaf from the smallholders as well as to reduce the production cost to the manufacturer. The payment schemes were introduced from late 1960s and there were various changes took place time to time and it was currently under the control of Tea Commissioner. The current formula was called as New Reasonable Price Formula (NRPF) which was prepared based on net sale average. Thus NSA of the factory was equal or less than the monthly elevational average price of made tea (E.Av), then smallholder suppliers were paid 68% and 32% were remain with manufacturer from the earnings. However, if the NSA was greater than monthly elevational average, then same income sharing was applicable up to E.Av and rest is divided among both parties equally as 50% [41]. The net outturn percentage of made tea to green leaf also has a greater impact on the pricing which was considered as 21.50% for low country made tea [50].

Economically Motivated Adulteration (EMA)

The motivation of EMA is financial, to gain an increased income from selling a foodstuff in a way which deceives customers and consumers. Current industry's one of the biggest threats to end the Ceylon tea production is the economically motivated adulteration. As one of main issues in the subcontinent, tea manufacturers in the country are illegally practicing the various EMA activities as a means of increasing the auction price. As the major EMA activity, there are evidence that many manufacturers use adulteration with addition sugar solutions to the rolling, some are usually apply dissolved concentrated sugar solution while some apply liquid glucose to improve the liquor colour. Earlier they also use ferrous salts to do the same which has been abandoned by now, but there are many manufacturers still practicing the sugar adulteration and they also claim that it is the colonial era practice where some of the factory owners want Sri Lanka Tea Board to legalize it. However, the country is already experiencing the highest COP in the world where this initiative will someday write the death certificate of Sri Lankan tea industry, if SLTB approve the proposal. Hence, it is better to explore the modes of EMA and their threats to the food industry as a nonfinancial barrier to exports.

The EMA may be by either passing off a cheaper material as a more expensive one, or it may be that a less expensive ingredient is used to replace or extend the more expensive one. The avoidance of loss may also be an incentive for adulteration. Limited supply of a key material may encourage a producer to improvise to complete an order rather than declare short delivery to the customer. The intention of EMA is not to cause illness or death, but that may be the result. This was the case in 2008 when melamine was used as a nitrogen source to fraudulently increase the measured protein content of milk, resulting in more than 50,000 babies hospitalized and six deaths after having consumed contaminated infant formula.

The common factor in many cases of EMA is that the adulterant is neither a food safety hazard, nor readily identified, as this would defeat the aim of the attacker. Common adulterants in tea industry include refuse tea, second grades and sugar; ingredients that may be properly used and declared but improper use is food fraud. EMA is likely to be more effective for an attacker, and therefore present a greater threat to a food business, upstream on the food supply chain close to manufacture of primary ingredients. A successful adulteration continues without detection.

Adulteration

To further explain the potential public health threats from EMA, and to develop appropriate mitigation plan, it is important to discuss adulteration methods. There are 8 major adulteration types that are; dilution, substitution, artificial enhancement, mislabeling, trans-shipment and origin masking, counterfeiting, theft and resale, and intentional distribution of contaminated product. Dilution, substitution and artificial enhancement involve altering the food itself. The other five methods involve altering the exterior appearance, misrepresentation and/or violating commercial regulations.

Dilution

Dilution is achieved through the addition of a cheaper ingredient to increase the overall weight or volume of a product. For example, most of tea available in the local market are adulterated with lower quality cut and artificially coloured and flavoured fiber material in the blends.

Substitution

Substitution is the replacement of an authentic product with a fraudulent one. A common example is the substitution of lower quality imported tea for premium products.

Artificial Enhancement

Artificial Enhancement enhances a product's attributes by use of an unapproved additive. The very famous adulteration in the country are addition of sugar syrup or dissolved sugar that impart the strong red colour in the brew. There are various other compound were used before, but most of them are abandoned except sugar currently.

Mislabeling

Mislabeling occurs when quality, harvesting, or processing techniques of a food product are misrepresented. Examples include selling Indian or Vietnam tea labeled as Ceylon tea, which was also practiced by many big entrepreneurs in the country time to time. There are various reported cases in the past.

Trans-shipment

Transshipment moves goods through an intermediary country before shipping to its final destination thereby masking its true origin. Motivations that include differences in regulatory oversight, consumer perception and avoidance of tariffs lead to transshipment. Transshipment can be an in-country issue as well.

Counterfeiting

Counterfeiting refers to the mimicking of one food by replicating it out of different ingredients which can be committed by fraudulently using a brand-name label on an inauthentic product, thus selling one product as a different — often more valuable — product.

Non-Financial Barriers in Tea Industry

Since Sri Lankan tea industry is highly export driven where 90% of the produce is exported. Growth of NGO movements and consumer campaigns in developed countries where tea is imported are demanding for social and environmental responsibilities throughout the supply chain starting from the plucking of green leaf to the end user. Thus buyers in the auction demand for such measures before purchasing tea from the local manufacturers with social policy containing detailed guidelines on labour situations-in the field of employment, conditions of work and industrial relations as well as recent UN human rights norms on the responsibilities ensuring the observation of human rights, right to equal opportunity and non-discriminatory treatment, rights of the workers, use of due diligence and promotion of economic, social and cultural rights. On the other hand, corporate social responsibility (CSR) initiatives such as

Ethical Tea Partnership (ETP) and Fair Trade certifications are gaining significant momentum in buyer decisions where there is scant information on the environmental, social and economic impact of tea production and processing in Sri Lanka [41].

These factors have huge weightage on cost of production as well as price achieved in the tea auction and that needs fairly high capital investment requirements as well as cost to the final product. Nevertheless, there is a severe worker shortage in plantations as well as processing which is also true for corporate sector and smallholder sector. The industry also lacks qualified and skilled staff and skilled labourers to supervise tea cultivation and processing as well as pruning, fertilizing and weeding where industry got single alternative which is to mechanize tea fields and factory operation in future. Hence fears are expressed, whether Sri Lanka could still maintain its image as the producer of best quality tea [41].

On top of that, current context of food factory concepts which needs to comply with basic hygienic requirements while certifying for voluntary certification systems such as HACCP, ISO 22000, FSSC 22000, ISO 9001, ISO 14001 and mandatory regulations such as SLTB monitoring measures has to be met. Thus manufacturers have to automate and modernize the tea factories to achieve such certifications where they need more capital infusion.

Alternatively, increased demand for social and ethical conditions (eg: Fair Trade labeling, ETP etc.,) has to be complied while requirements for very low levels of pesticide residues in made tea, especially for European Union (EU) Countries and Japan, had almost become a non-tariff barrier for export of tea from Sri Lanka [41].

However, Sri Lanka can be proud of, being the one and only tea exporting country, which strictly adheres to ISO 3720 standard (minimum product quality), for each kilogram of tea exported and there is consensus among the producers that, all tea factories should be developed to requirements set up as per Hazard Analysis and Critical Control Points (HACCP) or ISO 22000/FSSC 22000 standards. Consequently, these standards are very useful to ensure safety of the product manufactured; i.e. to ensure the product is free of physical, chemical and biological contaminants. Accordingly, almost all the factories have initiated action towards obtaining these certifications, if they have not achieved the same already. However, some plantation management companies and private bought leaf factories face difficulties in generating funds to meet the cost involved, as product/system certifications require improvements to product line, hygiene, welfare etc. [41].

5
Standard Environment in Tea Trade

After 2000, there is an overwhelming demand for the product standards which later shifted into process standards and it further tighten the application of public standards. Nonetheless, this has widened the scope of the standard while increasing the importance of the private standards which has created concerns in developing countries as they are becoming big block of non-tariff measures. Since WTO member countries should follow reduced tariff over imports, developed countries are using these measures to control imports from developing countries through standards, which intern could potentially result in loss of international market, a decrease in employment and a decline of an industry. Thus, compliance improves brand image, competitiveness, and opportunities in new markets which encourage the advancement of an industry.

Considering the Sri Lankan tea industry, it is also increasingly governed by strict and complex standards, which reflects the evolving trends in the standards environment globally. Thus, Sri Lankan tea manufacturers should comply with various local and international standards pertaining quality and food safety including mandatory regulations by the Sri Lanka Tea Board, which includes ISO 3720, product standards for black tea. In addition, the manufacturer must further comply with foreign matter, micro biological contamination, heavy metal and pesticide residue limits, which are specified and monitored by the SLTB.

Nevertheless, Hazard Analysis and Critical Control Point (HACCP) and ISO 22000 Food Safety Management System are also gaining the significance in the tea trade which is voluntary in the current context, but compliance to either HACCP or ISO 22000 is becoming necessary and considered to be mandatory standards soon. Considering the market competitiveness and consumer preferences, there are a number of private standards which have been voluntarily adopted by tea exporters, in addition to the mandatory and voluntary public standards. The private standards usually go beyond the minimum compliance levels of public standards which demands further compliance in the area of food quality and safety, including social and environmental concerns. The list includes FSSC 22000, BRC (British Retail Consortium) Global Standard, Organic, Fair Trade (FT), Ethical Tea Partnership (ETP), and Rainforest Alliance (RA) are the most common private standards which have been collectively set and monitored by third party agencies.

Further to that, there are private standards or private codes of conduct which are specific to individual buyers such as supermarkets (i.e., Japanese supermarkets) and fast food chains (i.e., McDonalds) who have their own set of standards and requirements. Considering the standard environment in tea trade, the emergence of private standards reflects a growing interest by buyers, and ultimately consumers, about the conditions under which tea is produced [51].

The Evolution

The concept of quality has existed for many years, though it's meaning has changed and evolved over time. In the early twentieth century, quality management meant inspecting products to ensure that they met specifications. In the 1940s, during World War II, quality became more statistical in nature. Statistical sampling techniques were used to evaluate quality, and quality control charts were used to monitor the production process. In the 1960s, with the help of so called "quality gurus," the concept took on a broader meaning. Quality began to be viewed as something that encompassed the entire organization, not only the production process. Since all functions were responsible for product quality and all shared the costs of poor quality, quality was seen as a concept that affected the entire organization.

The meaning of quality for businesses changed dramatically in the late 1970s. Before then quality was still viewed as something that needed to be inspected and corrected. However, in the 1970s and 1980s many U.S. industries lost market share to foreign competition. In the auto industry, manufacturers such as Toyota and Honda became major players. In the consumer goods market, companies such as Toshiba and Sony led the way. These foreign competitors were producing lower priced products with considerably higher quality.

To survive, companies had to make major changes in their quality programs. Many hired consultants and instituted quality training programs for their employees. A new concept of quality was emerging. One result is that quality began to have a strategic meaning. Today, successful companies understand that quality provides a competitive advantage. They put the customer first and define quality as meeting or exceeding customer expectations.

Since the 1970s, competition based on quality has grown in importance and has generated tremendous interest, concern, and enthusiasm. Companies in every line of business are focusing on improving quality in order to be more competitive. In many industries quality excellence, has become a standard for doing business. Companies that do not meet this standard simply will not survive. The importance of quality is demonstrated by many national quality awards and quality certifications that are coveted by businesses.

Evolution of ISO Standards

The quality assurance was always there from the prehistoric times, but it was not practiced as a concept at least until a century ago due to the misunderstanding that it was an additional cost to the product. International standardization began in the early 20th century with the creation of the International Electro Technical Commission (IEC) in 1906 [52] and the ISO was begun in 1926 under the name of International Federation of the National Standardizing Associations (ISA) which was basically focused on mechanical engineering and disbanded in 1942 during the Second World War. However, it was reorganized in 1946 as ISO which is the current name of the organization and it was stipulated as a voluntary umbrella organization where it's members were national standardization authorities; each member representing one country [53]. The practices spread around the world while quality standards were further developed where International Organization for Standardization (ISO) was formed with headquarters in Geneva, Switzerland. The ISO published its first standard in 1951 and it has published 10,060 standards by 1998 [52].

However, during the World War II quality assurance initiated as a primary concept in production to deal with quality issues identified in defense equipment and weaponry. Thus, U.S. Department of Defense initiated and implemented one of the world first quality control programs where finish product testing was the main concern. This initiative was followed by the United Kingdom who developed their own quality standard for military equipment manufacturing. These practices were later spread into other countries where it formed a basis for development of set of quality assurance standards which was adopted by the all members of North Atlantic Treaty Organization (NATO) and it was called the Allied Quality Assurance Publication (AQAP). The development continued even after the World War II and US government use the quality standards while contracting defense equipment purchasing where contractors were encouraged and expected to implement those standard practices within their manufacturing processes.

ISO Standards were not limited to one or two subject area, it covers a multitude of topics including, but not limited to, paper sizes, a uniform system of measurement, symbols for automobile controls, film speed codes, and an internationally standardized freight container, etc. Consequently, one of the initial members of the Organization; the British Standards Institute (BSI) initiated a proposal to develop an international quality assurance and management standard in 1979 [52]. In addition, UK government published a white paper on "Standards, Quality and International Competitiveness" that introduced the concept of achieving certification as a marketing tool in 1982 [54] which helps ISO standards to become as a popular marketing tool. Nonetheless, the ISO commenced a technical committee on the initiative which was ISO/TC 176 with twenty member nations and another 14 nations as observing members. Thus, ISO published its first quality assurance standard in 1987 after eight years of hard work from the initiation which was called the ISO 9000 series [52]. The initial ISO 9000 standard was mostly identical with BS 5750 with three models for quality management system and the selection was based on the scope of the organization. On the other hand, it reflects much of its military inheritance in the initial version [54]. The methodology and practices learnt in the process was applied in to other areas of manufacturing to help the industry as well as to support the globalization of human society where various ISO series were developed later, i.e. ISO 14001 Environmental Management System, ISO 27001 Information Security Management, ISO 22000 Food Safety Management System, etc.

ISO 9001:2015

ISO 9001:2008 was the previous quality standard applicable in the market which was further improved into ISO 9001:2015 recently. All the versions of the ISO 9001 had intended to evaluate the firm's ability to effectively design, produce, and deliver quality products and services while enhancing customer satisfaction by including more top-management involvement and continual improvement [55]. ISO 9001 use a process approach which aims to achieve customer satisfaction by meeting customer requirements while improving the system continuously and to prevent nonconformity in products and/or services. The standard provides guidelines for organizations to establish their quality systems by focusing on procedures, control, and documentation [56], while conceptualizing that certain minimum characteristics of a quality management system could be usefully standardized, giving mutual benefit to suppliers and customers, and focusing on process rather than product/service quality [57]. Considering the customer focus as one of the key area of customers' needs and expectations, one of the most important customer expectations in their list is to have safe food products, where ISO 9001 allows an organization to integrate its quality management system with the implementation of a food safety system [58]. According to Bolton [59], food companies can implement quality assurance systems according to ISO 9000 series, while ensuring quality procedures and reinforcing legislative requirements. On the other hand, ISO 9001 standard is internationally recognized and designed to demonstrate its ability to transform the supplying organization to achieve a basic level of quality by the formalization and documentation of its quality management system, where its effective deployment has been widely recognized in recent years as a means of building sustainable competitive advantage and thereby enhancing firm performance [60].

Food Safety and Standards

According to the Webster's Ninth New Collage Dictionary [61] safety was defined as the "condition of being safe from undergoing or causing hurt, injury or loss" and according to the FAO and WHO [62] food safety was the "assurance that food will not cause any harm to the consumer when prepared and/or consumed according to the intended use". Ensuring food safety in current complex society was an intimidating task which was possible only with corporation and collective efforts of all stakeholders in food supply chain including consumer organizations, industry and the government [63]. On the other hand, food safety was a global phenomenon growing its importance everyday due to the concerns in public health and impact on global trade [64]. Quality was an essential necessity of the competitiveness and organization's survival in highly competitive global economy with continuous improvement of product, process and services [65], [66] where industry has upward moving trends in implementing food quality assurance systems as well as food safety assurance systems.

The food safety assurance systems were required for manufacturing organizations to ensure food safety and compliance to statutory and regulatory requirements as well as customer requirements in food supply chain [67]. On the other hand, organization's competitiveness and position in global food trade can be strengthened through implementation of quality assurance systems [68] where quality management system can be defined as a complete set of written procedures, practical applications, records of evidence and training [69]. Food manufacturers were interested in implementing food safety and quality systems such as Hazard Analysis Critical Control Point (HACCP) and ISO 9001 Quality Management Systems to comply with quality practices [70] where ISO 9001 basically concentrate on process management requirements while HACCP is focused on technological aspects of food safety assurance [71]. Thus, Manning and Baines [72] emphasized that both food safety and quality of the product and its manufacturing process can be addressed through effective quality assurance systems by splitting product and the process where quality can be defined in terms of intrinsic quality (product) and extrinsic quality (process).

Accordingly, the ISO 22000: 2005 was introduced to world with objective of food safety in food supply chain [67] where synergetic effect of ISO 9001 and HACCP was expected instead of applying two systems in food industry. Thus, ISO 22000 has created a more resilient basis for establishing and demonstrating compliance of organization's quality management systems with appropriate documentation and procedures defined by the standard, where controls has to be established for every aspects of production process while documenting all the operational procedures as well as managerial actions [73].

ISO 22000 is developed for food industry where it is directly applicable with the core production areas of tea manufacturing process which is a valuable tool for tea manufacturers in order to ensure that both quality assurance standards and food quality procedures have been met and achieved. The implementation of ISO 22000 in tea industry is related to the structure of tea factory, to the nature and number of tea grades that produced and consumed globally and finally, to the procedures of production. It is well accepted that tea has an advantage of food safety due to the nature of tea manufacturing process where ISO 22000 certification procedures can serve to identify deficiencies in processes or quality controls from production to consumers. As a result tea manufacturers will ensure that the production will comply with quality and food safety standards.

Due to the growth of information age as well as creation of awareness among consumers, the food safety requirement is never been so as high today where ISO 22000 has become a valuable tool in assessing and preventing food safety even before it started. In contrast, quality is a very difficult term to define or to understand and measuring which cannot be taken as an absolute. The quality assurance is a guarantee of agreed-upon specifications has been delivered [73]. In addition to that, few writers conclude that, explicitly or implicitly that quality is simple, nevertheless many treatises on quality conclude that it is complex, multidimensional, and relative [74]. According to Juran [75], Quality is not a scientific or a technical word, it is not a physical entity, but it is a very useful concept in general life and management. Thus, terms "food quality" and "food safety" mean different things to different people based on their perception.

Consequently, food quality was considered as an interesting concept where it transcends all steps and all actors within the food chain covering one step forward and one step backward; however it is of an intangible nature because it is perceived individually [76]. Food quality meanings can be varying according to the situation and can encompass parameters as diverse as organoleptic characteristics, physical and functional properties, nutrient content and consumer protection from fraud. On the other hand, safety is more straightforward, relating to the content of various chemical and microbiological elements in food [77]. Food safety and food quality assurance are forms of guarantees, where assurance of quality is a guarantee that agreed-upon specifications have been met. However, if the safety related specifications are included in the quality assurance system, then the assurance of quality incorporates safety [78]. Nevertheless, consumer is the key to defining quality, where a company's internal definition of quality is meaningless if it fails to reflect consumer requirements [79].

In the current context, food firms are facing increasingly intense competitive markets and are implementing quality assurance systems [70] where each quality assurance system covers different quality aspects e.g., some focus on management aspects (ISO), whereas others focus on technology aspects (GMP, HACCP). The current standards are developed focusing to run on multiple platform quality assurance systems which are often combined to assure several quality aspects, for assuring food safety and food quality e.g., the combination of HACCP and ISO 9000 [80].

HACCP

The Hazard Analysis Critical Control Points (HACCP) system was first invented for NASA with collaboration of Pillsbury Corporation and US Army due to the risk involved in supplying preserved food for astronauts [81] and it is a common sense approach in identifying, quantifying and controlling food safety hazards.

The HACCP allow food manufacturer to carry out a detailed examination of a process to identify hazards and where the hazards can be controlled by setting up a framework [82] which is a food safety management strategy that has been widely tested and established as an effective means of preventing food-borne diseases when correctly implemented [83]. HACCP has designed in a way that it can be considered as a scientific and systematic system to assure food safety [84], while applying throughout the whole food chain [85], [86].

HACCP is a management system where food safety was addressed through the analysis and control of biological, chemical and physical hazards starting from raw material production, procurement and handling to manufacturing, distribution and consumption of the finished product at consumers end. The HACCP system is a proven, cost-effective method of maximizing food safety, where it focuses on hazard control at its source and consists of seven principles of international acceptance that outline how to establish, implement and maintain an HACCP plan for an operation under the consideration [87]. Nevertheless, most of the countries had made responsible food manufacturers to oblige by legislation to apply HACCP, while other systems are applied voluntarily in the food industry.

In addition, FDA has emphasized the role of prerequisite programs (PRPs) for the implementation of HACCP [88] where it has been recommended to apply prerequisite programs before the HACCP plan is utilized, [89]. Besides, HACCP complements the total quality management because it offers continuous problem prevention [90]. Accordingly, adaptation to a food quality/food safety management system and being able to signal it to consumers, firms can gain marketing advantage and competitive advantages in the consumer level [91].

ISO 22000

Significant food crises in world during the past decades have raised doubts in the consumer's mind and created a lack of trust and confidence in products placed on the market. Fortunately, most companies already take product quality and consumer safety very seriously. A lot of good practices have been developed and implemented on a voluntary basis. These practices ensure that product safety has never been as high as it is today [92]. Companies continuously challenge their internal quality systems and work on continuous improvement, thanks to new technologies and ways of working.

International trade of food products are increasing while increase in scientific knowledge about hazards associated with foods and their consequent effects on health have made people critically think about their food habits. Thus, there is a growing concern on food safety, because, growing consumer awareness, more foods prepared away from home, rising of incidence of food born illness in many countries, globalization and less barriers to trade present new food safety challenges, unfamiliar hazards or new hazards. For an example, 70% of the approximate 1.5 billion case of diarrhea that occur globally each year are directly caused by chemical or biological contamination of food and more food allergies have been reported over recent years, and the number of people with food allergies are gradually increasing [93].

Foodborne illness is a preventable disease affecting all people, which has significant impact on public health and significant trade implications on economies. As to the published data, around 76 million cases of foodborne illness occur each year in the United States, costing between $6.5 and $34.9 billion in medical care and lost productivity [94], [95].

To date, there are 250+ types of food borne illness have been identified with the effects ranging from acute to chronic illness such as mild symptoms to life threatening. Foodborne illness is significantly underreported, due to the lack of awareness among community where diarrhoeal diseases alone; a considerable proportion of which is foodborne illnesses that kills around 1.9 million children globally every year [96]. In addition, food born diseases cause 76 million illnesses while hospitalizing around 325,000 with 5,000 deaths in the United States each year [95].

Over 40 different food born microbial pathogens including fungi, viruses, parasites, and bacteria, were believed to cause human illnesses at the time and it was estimated that for six bacterial pathogens, the costs of human illness were estimated to be USD 9.3 – USD 12.9 billion annually, of these costs, USD 2.9 – USD 6.7 billion were attributed to food borne bacteria [97]. These estimates were developed to provide an analytical support for USDA's Hazard Analysis and Critical Control Point (HACCP) system initiated for meat and poultry at the beginning which overrule entire food industry today with amalgamating in to the core of the various global food standards.

Previously, product safety was perceived and positioned as the voluntary responsibility of companies but the publication of EU Directive 2001/95/EC on General Product Safety in December 2001, and EU Regulation 178/2002 on Food Safety in January 2002 brought about a significant change. Today, European legislation constitutes a set of requirements that each company manufacturing, distributing, importing and/or exporting products to and from Europe must comply with [92].

Beyond the legal aspect, consumer safety is primarily a question of business ethics and responsibility. Good product quality and product safety contribute to build up consumer confidence and consequently strengthen the image of a company or a brand in the consumer's mind. Failure to respect consumers' needs and expectations may be interpreted as betraying this confidence and consequently may lead, in the long term and the worst case, to damage for a company and its brand image and in some cases for the business partners and the whole industry. This is what is at stake when quality and safety are compromised.

Considering these food safety problems and trade issues generated over the time, the International Standard Organization developed the ISO 22000 Food Safety Management System to harmonize the requirements of various food safety standards into integrated food safety management system while eliminating lots of trade issues faced on exports. Thus ISO 22000 is an international, auditable standard which specifies the requirements for food safety management system by incorporating all the elements of Good Manufacturing Practices (GMP) and Hazard Analysis Critical Control Points (HACCP) together with a comprehensive management system [98]. The standard ensures the complete food safety of entire food supply chain while satisfying global food safety statutory and regulatory requirements. It promotes the conformity to the international standard of the product or services offered by providing the assurance of quality, safety and reliability [99].

According to the Food safety experts in the field, set of well-functioning prerequisite programs (PRPs) initially simplify and strengthen the HACCP plan, where ISO 22000:2005 was a HACCP-type standard which fits very well with ISO 9001:2000 because, it was especially developed to assure food safety. ISO 22000 has dynamically combine the HACCP principles and application steps with prerequisite programs, using the hazard analysis to determine the strategy to be used to ensure hazard control by combining the prerequisite programs and the HACCP plan [100]. Nevertheless, ISO 22000:2005 was the first in a family of standards which was entirely focused on food safety that introduced focusing entire food chain.

However, the several food safety standards evolve overtime, but most of them are based on the guidelines of ISO 22000 and HACCP, hence adhering to ISO 22000 usually covers minimum compliances which are required internationally. Hence, rest of the additions are basically applied in private standards such as FSSC 22000, etc. However, the ISO 22000 released its first revision in 2018, where the current applicable standard is ISO 22000:2018 and it consists of several sister standards which are usually applied together based on the specific requirements in the supply chain which includes the following documents:

- ☯ ISO/TS 22003:2013, Food safety management systems – Requirements for bodies providing audit and certification of food safety management systems.
- ☯ ISO/TS 22004:2014, Food safety management systems – Guidance on the application of ISO 22000.
- ☯ ISO 22005:2007, Traceability in the feed and food chain – General principles and basic requirement for system design and implementation.
- ☯ ISO 22002-I:2009, Prerequisite programs on food safety – Part I: Food manufacturing.

The ISO/TS 22002 series further specifies requirements and guidance for establishing, implementing and maintaining prerequisite programs (PRPs) to assist in controlling food safety hazards. These Technical Specifications are:
ISO/TS 22002-1 Prerequisite programs on food safety - Part 1: Food manufacturing
ISO/TS 22002-2 Prerequisite programs on food safety - Part 2: Catering
ISO/TS 22002-3 Prerequisite programs on food safety - Part 3: Farming
ISO/TS 22002-4 Prerequisite programs on food safety - Part 4: Food packaging manufacturing

ISO/TS 22002-6 Prerequisite programs on food safety - Part 6: Feed and animal food production.

Hence, the international standard specifies the requirements for a food safety management system (FSMS) that combines the following generally recognized key elements to ensure food safety along the food chain, up to the point of final consumption, that are,

1. Interactive communication
2. System management
3. HACCP principles
4. Prerequisite programs

In addition, the standard is also based on the principles that are common to ISO management system standards, particularly the Annex SL format or high level structure which are;

5. Customer focus
6. Leadership
7. Engagement of people
8. Process approach
9. Improvement
10. Evidence-based decision making
11. Relationship management

As ISO 22000:2018 explains, the process approach of the standard involves systematic definition and management of processes, and their interactions, to achieve the intended results in accordance with the food safety policy and strategic direction of the organization. Hence, management of the processes and the system can be achieved using the PDCA cycle as a whole with an overall focus on risk-based thinking aimed at taking advantage of opportunities and preventing undesirable results, because understanding and managing interrelated processes as a system contributes to the organization's effectiveness and efficiency in achieving its intended results. Thus, communication along the food chain is essential to ensure that all relevant food safety hazards are identified and adequately controlled at each step within the food chain. This implies the importance of communication between organizations between both upstream and downstream in the food chain. Nonetheless, recognition of the organization's role and the position within the food chain is essential to ensure effective interactive communication throughout chain in order to deliver safe food product to the end user [101-a].

Further, ISO 22000:2018 has expanded the use of PDCA (plan-do-check-act) cycle (figure 5.1) while integrating HACCP into the common cycle used in the ISO 9001 which has been segregated to two levels as organizational planning and control and operational planning and control. Hence, first covers the overall frame of the FSMS (Clause 4 to Clause 7 and Clause 9 to Clause 10), while operational planning and control covers the operational processes within the food safety system as described in Clause 8, where communication between two levels is essential. Here is the brief explanation given by the standard regards to its new expansion [101-b].

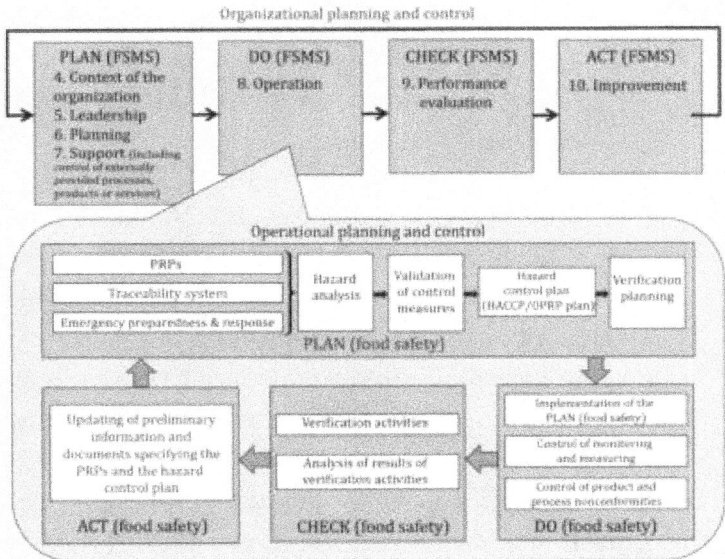

Figure 5.1 – ISO 22000:2018 PDCA cycle
Source: ISO_DIS_22000_(E)

Plan:
Establish the objectives of the system and its processes, provide the resources needed to deliver the results, and identify and address risks and opportunities;
Do:
Implement what was planned;

Check:

Monitor and (where relevant) measure processes and the resulting products and services, analyze and evaluate information and data from monitoring, measuring and verification activities, and report the results;

Act:

Take actions to improve performance, as necessary [101-b].

The new version integrates PDCA in a more comprehensive manner where planning has absorbed the HACCP system and application steps developed by Codex Alimentarius Commission; by means of auditable requirements, it combines the HACCP plan with (PRPs) perquisite programs, traceability system and emergency preparedness and response through operational planning and control. On the other hand, standard offers an alternative to food manufacturers who do not implement ISO 9001, while they want to have an effective food safety management system [58] as it combines a series of advantages, involving quality management, external and in-house communications, designating responsibility, implementing crisis management, continual improvement, good health practices and differentiating between PRP, OPRP and CCP [102].

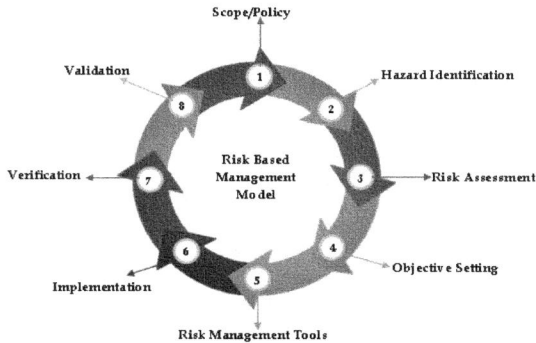

Figure 5.2: Risk based management model (RBMM)

The ISO 22000 FSMS has been developed based on risk based management model focusing the entire food supply chain through harmonization. The risk based management model has eight steps. Through RBMM, each and every processing step is evaluated for its suitability, if any step is not complying with validation requirements, (figure 5.2) will start from the beginning until it can be validated.

The 3 major pillars/layers of ISO 22000 FSMS has not changed from its initial debut in the new version; they can be shown below figure 5.3, where ISO 22000 has been developed basically merging GMP, HACCP and ISO 9001 as to the initial development of the standard. Hence, core of the standard is still the same, even though there are several new changes has been proposed. As to the figure 5.3, the foundation layer is consist of GMP/GHP/GAP, Codex General Principles of

Food Hygiene and Prerequisite programs which altogether creates very sound infrastructure and physical requirements to implement food safety requirements inside the plant focusing on basic food hygiene standards.

Figure 5.3 – Three layer model of ISO 22000:2005

The total food safety is achieved through HACCP system of Codex Alimentarius while using its seven principles indirectly in different terminologies to identify hazards and to control them under strict management plan (figure 5.1). This includes the hazard analysis, identification of critical control points, establishment of critical control limits, monitoring procedures, corrective actions, record keeping, validation and verification activities. However, the use of HACCP word has been minimized to the bare minimum while promoting the hazard analysis.

The major changes in the standard applied in documentation process, since Annex SL format has adjusted its wording from documentation and record keeping to documented information in the recently modified versions of ISO standards. Hence, at the documentation level, the standard has flatten its controls while introducing the bare minimum thus, releasing the complete responsibility to the user, which is one of the best improvements, because standard 3rd party auditors are crazy and hectic headache to users. They change their documentation requirements from one audit to another, person to person, organization to organization if they couldn't find any minor concerns, but they mostly ignored major concerns to retain the client. Now company can decide what documents to be kept, at which format, the way they want to document, and what documents are really required based on the process requirements, but it doesn't mean that company do not require documentation. Hence, there are mandatory documented information as evidence of proper implementation and maintenance, which is mentioned in the standard without basically dictating whether it is a manual or a procedure or a work instruction or something else.

However, considering the documented information, it still requires much more than it asked from the food manufacturer where it is best advised to consider the intimal documentation model or what we call the documentation pyramid to ease the development of better FSMS. The company can eliminate the manual, but manual can be used as an explanatory document of company food safety/quality system which can amalgamate several ISO standards together such as ISO 9001, ISO 22000, ISO 14001, etc. Thus, common Annex SL can be used to setup initial documents and references to the relevant procedures or secondary documents as well as system performance documents, while linking several standard together to build a single company manual. Nonetheless, it also provide the opportunity to club all the relevant documents to a one place while using them in a flattened structure.

Considering the mandatory documented information, the standard still requests to provide hazard analysis, PRPs, CCPs, OPRPs, hazard control plans, and procedures or sort of evidence of protocols for the below mentioned areas. In addition, same procedures and activities are applied to the prerequisite programs and operational prerequisite programs identified according to the risk levels of the product manufactured.

The ISO 22000 management elements are handled through mandatory documented information in addition to the above mentioned areas which consists of;

    1. Control of documented information

2. Statutory, regulatory and customer requirements
3. Communication
4. Product identification and traceability
5. Emergency preparedness and response
6. Corrections and corrective actions
7. Handling of potentially unsafe products
8. Withdrawals and recalls
9. Internal audits
10. Management review

Some of these procedures are not requested as procedures directly, but as documented information, but practitioners of ISO 22000 can consider to write them as procedures since it is easier to upgrade from previous model to the new version without much documentation work rather than revising the existing system. Nonetheless, some of these requirements are basically identical to ISO 9001, and compatible with its requirements, thus it can be used in harmony if required. The ISO 22000 FSMS has procedure for emergency preparedness and response, which is inherited from reputed safety standards while it is also identical to ISO 9001. The organization and the top management must be prepared to respond to potential emergency situations and accidents that can impact on food safety. These can include incidents such as fire, flooding, bio-terrorism and sabotage, energy failure, vehicle accidents, contamination of the environment, various types of weather-related events, or the impact of a pandemic [103].

Hence, food safety management system needs to be documented as it requires documented information. This means that the organization must have, as a minimum, a written food safety policy and related objectives, the procedures/way of execution/rules and performance evidence as required by ISO 22000 as well as any other documented information that might need to ensure the effective development, implementation and updating of the system. In addition, organization will not only need to document its policies and procedures but it also need to have a procedure for controlling its documentation and records, because food safety management systems will change over time, as will the people doing the activity. Therefore, one reason for controlling documented information is to ensure that the individual using the document has the most recent version of the document. Part of document control ensures that all the proposed changes are reviewed prior to implementation which determines their effects on food safety and the impact on the management system.

The documentation system is still identical to the ISO 9001 which consisted of four layers [104] in its previous version that has been flatten in the latest revision in 2015. As the organization develops its food safety management system, it has been advised to carefully document its activities. These will include the written food safety policy and related objectives, food safety procedures and the required records as to pervious terminology, even though their names have been changed to a generic documented information. Further, the scope of the required documentation has been extended wherever possible. For example, in establishing your control measures you are required to document your hazard analysis and hazard assessment, including the decision-making process and the selection of control measures, whereas decision making process requires additional categorizations based on two types of logics which require more information than previous version. Nonetheless, organization will have to have evidence of documented information on the validation of its system and verification activities. The work of the food safety team and the management review also require documentation as it was required before.

Prerequisite programs were basically developed as part of good manufacturing practices initially and later-on it was became one of the major components in HACCP, because most of the system developers wanted to keep lowest number of HACCP studies in a system where PRPs were used to cover less critical control points as well as which cannot be measured in real time. In ISO 22000:2005, this uncertainty was addressed with separating real time immeasurable critical control points in to operational prerequisite programs, which was not properly segregated in HACCP, even though later versions of HACCP addressed the issue up to a certain extent. As it didn't completely cover the gap until the ISO 22000:2005 was released, where ISO 22000:2005 version introduced separation through ISO 22000 decision tree with logical sequence of questions to segregate them. However, most of the users never understand the complete logic behind it and they compel to use HACCP decision tree, where there was a great disagreement between many auditors and user to clearly define them. Because, people love logical trees and select CCPs easily, but distinguishing OPRP had great differences which was ignorant in many cases.

Considering the changes in the new version of ISO 22000, it has nominated HACCP plan and OPRP plan as hazard control plan by clubbing both CCP plan and OPRP plan together which is a terminological improvement, rather than system, where standard has struggled to come out of better solution to differentiate, segregation of CCP and OPRP. But the technical committee has not come out with a completely successful solution which practitioners can easily understand. This is one of the major weak points so far in the previous standard, which is still remains a mystery to average users. When considering the private standards like FSSC 22000, it is also depend on the ISO 22000, where they have added additional parameters or streamlined the issues with the ISO 22000, but never address this part of the misunderstanding to the average user completely.

Nevertheless, all prerequisite programs have four common factors which are; address indirect or less critical food safety issues, cover general programs related to food safety and it can be applied to multiple production lines. Momentary failure to meet prerequisite programs seldom results in a food safety hazard [105]. The organization should use documented information of external origin relevant for food safety in its various activities, for example in meeting statutory, regulatory and customer requirements and their interests. The new version has also considered paperless management situations, where electronic documentation has added as part of documented information which may be required to comply with regulatory requirements.

As a whole, ISO 22000 can be considered as a business management tool which links food safety to business processes and encourages organizations to analyze requirements of interested parties, define processes and keep them in control where it enables integration of quality management and food safety management. In this way ISO 22000 FSMS is considered as more focused, more coherent and integrated food safety management system which can satisfy any food safety statutory or regulatory requirements.

Annex SL

Since ISO has hundreds of different standards, there was a requirement for common platform to reduce merging issues as well as updating them in a regular manner where ISO developed a common platform in few years back. The objective was to provide identical structure, text and common terms and definitions for management systems standards of the modern tech savvy global economy. The platform, known as Annex SL, or the high level structure was designed to ensure consistency among future and revised management systems standards, while making the standards easier to read and be understood by users, and greatly aid with the integration of multiple standards within one organization.

Annex SL is promised to be a new approach to management system format that helps streamlining the creation of new standards, and implementation of multiple standards within one organization easier. The Annex SL replaces ISO's Guide 83, which previously provided a base structure and standardized texts for management system standards (MSS). Initially, guide 83 was started to address complaints that many have received when integrating MSS like ISO 9001, ISO 14001, ISO 22000, and ISO 2700, etc. In addition, Annex SL also addresses much of criticism expressed by organizations integrating multiple management systems, because these standards have common elements, which are described and organized differently, making it difficult for organizations to implement multiple standards.

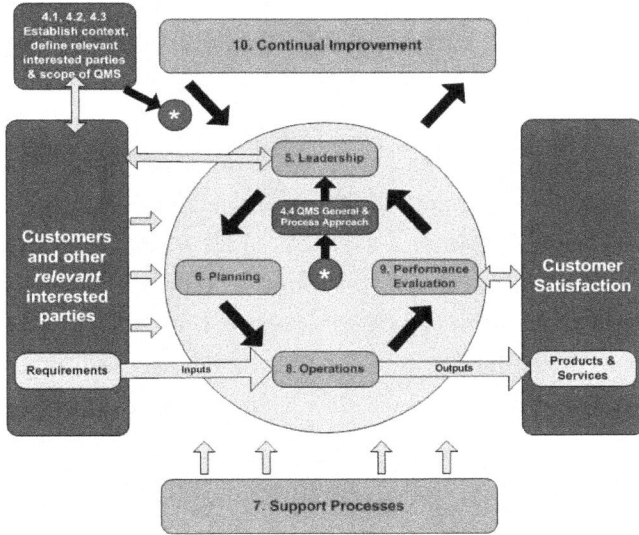

Figure 5.4: Annex SL process flow
Image courtesy: 3.bp.blogspot.com

Thus, Annex SL addresses these issues by creating a template upon which ISO MSS are to be built, where it was primarily developed as a guide to those who draft the standards. Hence, core of Annex SL consists of 8 clauses and 4 appendices that encompass a "high level structure" (essentially shared high level concepts among standards), shared terms/definitions and actual shared clause titles and text. The appendix three is in three parts which are high-level structure, identical core text and common terms and core definitions. Despite the fact that, these standards have common elements, they are described and organized differently, making effective integration difficult. Hence, development of Annex SL has improved the use of the same structure, as well as commonly used terms and definitions, will make it far easier, less time-consuming, and consequently cheaper to implement, integrate, and maintain standards. Nonetheless, Annex SL has enabled organizations to enhance alignment among ISO's management system standards, while facilitating their implementation for organizations that may need to simultaneously meet the requirement of two or more such standards. As a result of application of Annex SL in the development of new standards, approximately 30% or so of each new and revised standard will contain identical text.

The intent of the initiation of Annex SL to create all management system standards to have the same overall look and feel, where all standards will progressively migrate during their next phase. In addition, Annex SL describes the framework for a generic management system, but it require the addition of discipline-specific requirements to make fully functional standards for systems such as quality, environmental, service management, food safety, business continuity, information security and energy management.

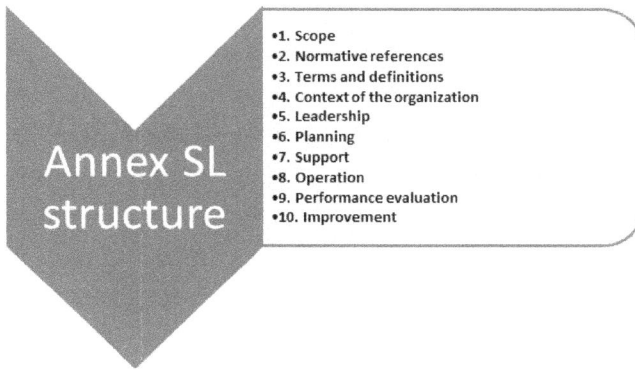

Figure 5.5: Annex SL Clauses
Image courtesy: 3.bp.blogspot.com

For MSSs writers, Annex SL provides the template for their work, which promote concentration of their development efforts on the discipline-specific requirements of the given MSS that focused on clause 8 – Operation. For management system implementers, Annex SL provides an overall management system framework within which they can pick and choose what discipline-specific standards they need to include, which will minimize the conflicts and duplication, confusion and misunderstanding from different MSSs. For management system auditors, there will be a core set of generic requirements that need to be addressed for all audits, no matter what discipline to be audited. Nonetheless, it could drive the development of auditor training, by addressing the common core set of requirements with additional training for discipline-specific requirements.

Thus, major clause numbers and titles of all ISO management system standards will be identical, such as the introduction, terms and definitions and operation, where introduction, scope and normative references will have content which is specific to each discipline and each standard can have its own bibliography. Overall, there is a reorganizing of management system requirements into this structure that may be unfamiliar, but some of the management system standards have already successfully migrated to this new structure such as ISO 9001, ISO 22000, etc.

Tea Standards

Tea was in the market from colonial era where there were many issues taken in to consideration and standards were developed accordingly. However, the oldest known modern standard is SLS 135 which was initially developed and published in 1979. The ISO 3720 was initially published in 1986 where Sri Lanka was the chair for the technical committee due to the expertise shown and the research and quality infrastructure available for tea manufacturing during the era. Other countries were far behind the quality at the time, but manufacturers didn't use it for the advantage where other immerging countries took the advantage of the new standards developed.

There are various tea standards but all of them are country specific other than the ISO 3720.

ISO 3720:2011

ISO 3720 was introduced in 1986 for the tea industry specially targeting the black tea, which was prepared by Technical Committee ISO/TC 34, Food products, Subcommittee SC 8, Tea and the latest version is 4th edition called as ISO 3720:2011 [105]. The previous versions were ISO 3720:1986, ISO 3720:1992 and ISO 3720:2004. However, tea is grown and manufactured in various parts of the world where desired characteristics of a black tea and the resulting liquor depend on many factors specific to consumer preferences and brewing methods where ISO 3720 standard set requirements for plant source from which black tea should be manufactured as well as set requirements for certain chemical characteristics as an indication of the tea, that can be used to recognized good production practice. According to the standard, the objectives are to "specify the plant source from which black tea is to be manufactured and to set requirements for certain chemical characteristics which, if met, are an indication that the tea has been subjected to recognized good production practice". The standard further explains that the quality of teas is usually assessed by tea tasters, and their judgments based on previous experience of tea from the specific producing area and their knowledge of national or regional conditions, and preferences in the consuming country. The standard also specifies the packing and marking requirements for black tea in containers. However, ISO 3720 is not applicable to scented or decaffeinated black tea.

The considerations for tea quality is based on characteristics such as the appearance of the tea before preparation of a liquor, the appearance of the infused leaf and the appearance, odour and taste of the liquor. However, ISO 3720 standard permits to depend on an expert tea taster who can assess whether a tea would be unlikely to comply with the chemical requirements or not, where time and expense can be saved by submitting teas for chemical analysis only if the tea is considered "suspect" by a tea taster as a general practice. Thus it is usually a requirement of the parties concerned whether to apply the requirements of ISO 3720 to a consignment or lot of black tea.

The application of the standard is voluntary in nature, but this standard is considered as the minimum requirements for export where quality of the tea was usually assessed by tea tasters, who base their judgments on own previous experiences of tea from the producing area and their knowledge of national or regional conditions, and preferences in the consuming country. There are several normative references to the test methods and relevant practices such as ISO 1572, Tea — Preparation of ground sample of known dry matter content,

ISO 1573, Tea — Determination of loss in mass at 103 °C, ISO 1575, Tea — Determination of total ash, ISO 1576, Tea — Determination of water-soluble ash and water-insoluble ash, ISO 1577, Tea — Determination of acid-insoluble ash, ISO 1578, Tea — Determination of alkalinity of water-soluble ash, ISO 1839, Tea — Sampling, ISO 3103, Tea — Preparation of liquor for use in sensory tests, ISO 5498, Agricultural food products — Determination of crude fibre content — General method, ISO 6078, Black tea — Vocabulary, ISO 9768, Tea — Determination of water extract, ISO 14502-1, Determination of substances characteristic of green and black tea — Part 1: Content of total polyphenols in tea — Colorimetric method using Folin-Ciocalteu reagent and ISO 15598, Tea — Determination of crude fibre content [105].

SLS Standards
Sri Lanka Standard Institution (SLSI) is the sole owner of Sri Lanka Standards or SLS trade mark and it is the official country representative for the International Organization for Standardization (ISO) which is a government institute. The SLSI make local standards and audit them and it also represent international standards for ISO officially. Mainly the SLSI has created two of them; where one was guideline while other was the standard. The SLS 135 Specifications for Black Tea for the local tea industry to maintain adequate quality with relevant

testing methods as well as auditing regularly. The standard initially launched in 1979 as local standard and mainly stresses the tea testing criteria for export which revised in 1986 and 2009. The most current one was named as Sri Lanka Standard SLS 135:2009 Specifications for Black Tea, which has designed to meet specific needs to guarantee the product quality while eliminating the adulteration. The standard was almost equivalent to ISO 3720 and it has all the testing criteria with test methods. SLS 1315: 2007; Code of Practice for Tea Industry - Part 1& Part 2, was the guideline document which use to create enabling environment to manufacture the product according to the SLS 135.

Product Certification Scheme for Tea (PCST)
There are lots of rejects and consumer complaints on Ceylon Tea, where SLTB took control the situation with its dictatorship over the industry due to government interests on the product. The SLTB and SLS jointly introduce a product certification scheme to certify the manufacturing process and the final product, i.e. Black Tea where the scheme was based on applicable standards of SLSI; Sri Lanka Standard SLS 135:2009; Specifications for black tea and ISO 3720:1986; Black tea- Definition and basic requirement as well as applicable SLTB regulations. The scheme called SLSI – SLTB: Product Certification Scheme for Tea (PCST) which was designed to meet the specific requirements of the tea plantation sector while providing internationally

recognized third party compliance to consumers. SLSI - SLTB Product Certification for Tea (PCST) is therefore comprehensive and it includes the following;

1. Applicable Tea Board Regulations & Sri Lanka Tea Board Standards/ Guidelines for Sri Lankan Origin Tea.
2. Sri Lanka Standard SLS 135:2009; Specifications for black tea and ISO 3720:1986; Black tea- Definition
3. Industry related code of Practices: SLS 1315: 2007; Code of practice for Tea Industry - Part 1& Part 2.

The given scheme was essentially voluntary in nature where it is largely based on ISO/ IEC Guide 65: 1996; General requirements for bodies operating product certification systems. The ISO/ IEC Guide 65: 1996 provides general rules for third party certification of determining with standards through initial testing and assessment of a factory quality management system and its acceptance followed by surveillance that takes into account the factory quality management system (QMS) and the testing of sample from the factory and open market [106].

SLTB Standards and Regulations
The Sri Lanka Tea Board was established in 1976 as a fully owned government statutory body for the development and control of tea industry. The SLTB is under the purview of ministry of plantation

industries where it's mandate is the regulation of production, cultivation and replanting, establishment and operation of tea factories and the conduct of the Colombo Tea Auctions. Nevertheless, its statutory responsibilities also include maintenance of tea quality standards, issuing packaging guidelines, warehousing requirements, etc., framed both under the Sri Lanka Tea Board Law and the Tea Control Act No. 51 of 1957 and the Tea (Tax and Control of Exports) Act No. 16 of 1959 [107]. The SLTB has created enabling environment to control quality of Sri Lankan origin teas by teaming up with SLSI.

The Sri Lankan tea is highly regulated by the SLTB with regulations developed through ISO 3720:2011, SLSI 135 and the SLTB Guidelines at the point of export or import because SLTB has the monitoring authority of quality/standards of teas exported is mandated to protect the image of Ceylon tea established globally. SLTB has introduced set of standards/guidelines in 2008 which directly impose the quantitative values which also call the minimum/maximum values to be maintained at the point of export or import. These values are further segregated into two different types based on the origin of tea which directly control the contaminants and adulterants with the stringent guidelines for "Sri Lankan Origin Teas" and "Other Origin Tea" which are imported for re-export purposes. However, the minimum quality standard still will be ISO 3720 black Tea which is mandatory at the point of export. These guidelines are in force due to the complaints received from various countries regarding the adulterants and contaminants in the black tea exported from Sri Lanka.

These standards and guidelines are explained in the regulation "Sri Lanka Tea Board Standards/ Guidelines for Tea Sri Lanka Tea Board Law No. 14 of 1975 Tea Control Act No 51 of 1957 Tea (Tax & Control of Export) Act No 16 of 1959 Sri Lanka Tea Board (Import & Export) Regulations 1981" which is issued in Circular No.: AL/MQS-Rev/2010 26th April 2010 by the SLTB with Annexure – A for the Sri Lankan Origin teas, other than established formats (i.e. Chemical contaminants, Extraneous (Foreign) matters, Microbial contaminants and Maximum Residue Limits of Pesticides etc.,) should be taken only as guidelines for compliance by all registered manufacturers and exporters of Sri Lankan Origin Tea. The Annexure – B is given for Other Origin Teas [107].

Specifications for Chemical Quality of Tea
The tea is highly monitored where all above explained standards are harmonized in a way to provide unmatched quality for the export market through land mark branding. The Sri Lankan teas are sold at highest prices in the market due to flavour characteristics, liquor colour and seasonal variations as well as differentiation of elevations.

But reality is that our COP is the highest in the world where it is not possible for the buyer to go below the fixed prices in the auction. Thus, regulatory authorities trying to provide best product with pure quality unmatched to other competitors in the market place. Thus following chemical, microbiological and physical parameters are set as quality parameters in PCBT scheme which is extracted from SLTB Circular No.: AL/MQS-Rev/2010 [107].

Table 5.1: Tea quality parameters

Name of the Standard	Acceptable Limit	Test Method Ref:
Water Extract	Min 32 % (m/m)	ISO 9768:1994
Total Ash	Min. 4% – Max. 8% (m/m)	ISO 1575:1987
Water soluble ash of total Ash	Min. 45% (m/m)	ISO 1576:1988
Alkalinity of water soluble Ash (as KOH)	Min. 1.0% – Max. 3.0% (m/m)	ISO 1578:1975
Acid insoluble Ash	Max. 1.0% (m/m)	ISO 1577:1987
Crude Fibre	Max. 16.5% (m/m)	ISO 15598:1999

Document Courtesy: SLTB Circular No.: AL/MQS-Rev/2010.

Sri Lanka Tea Board Standards for "Sri Lankan Origin Teas" (Annexure A), Basic Requirements for Black Tea: ISO 3720: 1986/ Corrigendum 1: 1992 (E) & 2:2004 (E) and/or Sri Lanka Standards 135:1979 (UDC 663.951).

Sri Lanka Tea Board Guidelines
1) Foreign Matter – Completely free (Teas should comply with ISO 3720:1986 parameters specified above)

2) Heavy Metals

Table 5.2: Heavy metals

Name of the Metal	Acceptable Limit	Test Method Ref:
Iron	Max. 500 mg/kg	AOAC: 975.03 (2007)
Copper	Max. 100 mg/kg	AOAC: 971.20 (2007)
Lead	Max. 2 mg/kg	AOAC: 972.25 (2007)
Zinc	Max. 100 mg/kg	AOAC: 969.32 (2007)
Cadmium	Max. 0.2 mg/kg	AOAC: 973.34 (2007)

(AOAC – Association of Official Analytical Chemists)

3) Microbiological Requirement

Table 5.3: Microbial quality

Name of the Standard	Acceptable Limit	Test Method Ref:
Aerobic Plate Count	Max. 10,000 cfu/g	SLS 516: Part 1: 1991/ISO 4833:2003
Yeast and Mould	Max. 10,000 cfu/g	SLS 516: Part 2: 1991/ISO 21527-2:2008
Total Coliforms	Max. 10 MNP/g	SLS 516: Part 3: 1982/ISO 4831:2006
E. Coli	Absent/g	SLS 516: Part 3: 1982/ISO 7251:2005
Salmonella	Absent/25 g	SLS 516: Part 5: 1992/ISO 6579:2002

4) Pesticide Residues

There is a given list of 27 pesticides which are recommended for application by the Tea Research Institute of Sri Lanka (TRI), where Sri Lankan Origin Teas should not contain residues of pesticides other than in this list. In addition, exporters of tea should be guided by the standards in the destination country as required by the importer.

Table 5.4: Pesticide residue limits

Name of the Pesticide	Category/ Type	MRL: EU Regulation (EC) No. 396/2005 (23/06/09) mg/kg (ppm)	MRL: Japan - Positive List (MHLW – May 2009) mg/kg (ppm)
2,4-D	W	0.1	NL (0.01)
Azadirachtin(Neeme xtract)	A/I/N	0.01	Exempted
Bitertanol	F	0.1	0.1
Carbofuran	I/N	0.05	0.2
Carbosulfan	I	0.1	0.1
Chlorfluazuron	I	0.01	10
Copper hydroxide	F	40 (as Cu)	Exempted
Copper oxide	F	40 (as Cu)	Exempted
Copper oxychloride	F	40 (as Cu)	Exempted
Dazomet	N	0.02	0.1
Diazinon	I	0.02	0.1
Diuron	W	0.1	1.0
Fenamiphos (Ph)	N	0.05	0.05
Fenthion	I	0.1	NL (0.01)
Glufosinate-ammonium	W	0.1	0.5
Glyphosate	W	2.0	1.0
Hexaconazole	F	0.05	0.05
Imidacloprid	I	0.05	0.05
MCPA	W	0.1	NL (0.01)
Metam Sodium	I/N	0.02	0.1
Oxyfluorfen	W	0.05	NL (0.01)
Paraquat	W	0.05	0.3
Propargite	A	5.0	5.0
Propiconazole	F	0.1	0.1
Sulphur	A	5.0	Exempted
Tebuconazole	F	0.05	30.0
Tebufenozide	I	0.1	25.0

A: Acaricide; F: Fungicide; I: Insecticide; N: Nematicide; W: Weedicides;

NL*: Denotes that MRLs are not established. However, default MRL of 0.01mg/kg is defined as the Limit.

Document Courtesy: SLTB Circular No.: AL/MQS-Rev/2010.

Maximum Residue Limits (MRLs) in EU and Japan for the respective pesticides are given for information.

The set of guidelines issued in the given circular has also mentioned the specifications for "Other origin teas" which is applicable when teas are imported to the country for re-export with the purpose of blending and mixing with local teas which affect the quality of local tea in the international market. Thus it has separate more or less similar set of guidelines under "Annexure B" of the above circular [107].

BRC Standard

British Retail Consortium introduce the BRC Food Technical Standard in 1998 to be used to evaluate manufacturers of retailers own brand food products, as a result of industry needs due to food safety issues in the market. The standard is designed to assist retailers and brand owners who produce food products of consistent safety and quality through contract manufacturers to assist their 'due diligence' defense, in case should they be subject to a prosecution by the enforcement authorities. The standard was originally developed in response to the needs of UK members of the British Retail Consortium, but it has gained usage world-wide and are specified by growing numbers of retailers and branded manufacturers in the EU, North America and other parts of the world. The certification is achieved through audit by third party certification service providers which reassure retailers and branded manufacturers of the capability and competence of the supplier. The independent third party compliance reduces the need for retailers and manufacturers to carry out their own audits, thereby reducing the administrative burden on both the supplier and the customer. The Certification of BRC applicable to products that have been manufactured or prepared only at the site where the evaluation has taken which further includes storage facilities that are under the direct control of the production site management.

The standard is usually updated regularly where most current version for food is "Issue 07" (BRC Global Standard for Food Safety Issue 7 was published in January 2015) which sets out requirements to specify the food safety, quality and operational criteria required to be in place within a food manufacturing organization to fulfill obligations with regard to legal compliance and protection of the consumer. The Standard is designed to allow an assessment of a company's premises, operational systems and procedures by a competent third party, the certification body, against the requirements of the Standard.

The major requirements of the BRC is also same as ISO 22000 and FSSC 22000 where adoption and implementation of HACCP, with an effective documented food safety/quality management system while controlling the factory environmental standards, products, processes and personnel.

The latest issue of this BRC Global Standard focuses on:

- ❧ Ensuring consistency of the audit process
- ❧ Providing a Standard that is flexible enough to allow extra voluntary modules to reduce the audit burden
- ❧ Encouraging systems to reduce exposure to fraud
- ❧ Promoting greater transparency and traceability in the supply chain
- ❧ Encouraging adoption of the standard in small sites and facilities where processes are still in development

Additionally, the issue 7 has focused on the audit towards the implementation of good manufacturing practices within production while adding increased emphasis on areas which usually have the highest rate of recalls and withdrawals, such as labeling and packing. Issue 7 further highlights the management commitment, Hazard Analysis and Critical Control Point (HACCP)-based food safety programs which also focus on supporting the quality management systems implemented. However, if you consider the structure and applications, BRC also have similar requirements as ISO 22000, but there are slight changes since it is a private standard which is usually requesting more than regulatory standards that usually requires minimum requirements. If you consider the food safety which is the most important part of the standard is HACCP, where there is no any difference to ISO 22000 or FSSC 22000. The prerequisite programs are basically GMP where there is no difference again for the foundation and food safety, but there are slight changes in management as usual.

FSSC 22000

According to Food Safety System Certification 22000 (FSSC 22000), the scheme is developed using the international and independent standards ISO 22000, ISO 22003 and technical specifications for sector PRPs, such as ISO 22002-1 and PAS 223, to design a more focused food safety management system which is a private standard that requires more than the minimum requirements to comply and it was developed through a wide and open consultation with a large number of related organizations. The FSSC 22000 certification scheme is developed as an independent private standard in response to the need of the international food sector to have an independent ISO-based food safety management scheme with third party auditing and certification. The FSSC 22000 certification is supported by the European Food and Drink Association (CIAA) and the American Groceries Manufacturing Association (GMA). FSSC 22000 is fully recognized by the Global Food Safety Initiative (GFSI) and Accreditation

Bodies around the world. According to the global food safety initiative (GFSI), ISO 22000:2005 was having problems in defining prerequisite programs because set specifications are not adequate to define PRPs, where GFSI has introduced clearly defined PRPs and other regulatory controls in FSSC 22000.

Considering the FSSC 22000, Food safety management and HACCP are greater part of the requirements which are directly based on ISO 22000, but the PRPs are having different perspective for ensuring food safety is that food manufacturing organizations maintain the conditions for hygienic environment and production. Since ISO 22000 has not fulfilled this requirement properly at the beginning, FSSC 22000 has taken explicit measures on the requirements of PRPs which allows ISO 22000 certification schemes to be benchmarked by independent regulatory organizations such as Global Food Safety Initiative of Foundation of CIES (GFSI), the British Standards Institution (BSI), the Confederation of the food and drink industries of the EU (CIAA).

The GFSI scheme recognized PRP requirements were developed in compliance with the general principles of food hygiene of the Codex Alimentarius, the guidelines for drinking-water quality of the WHO, the specification BSI-PAS 96 and the key element of good manufacturing practices of the GFSI guidance document (fifth edition; section 6.2). The developed standard also called the BSI-PAS 220, which considered most suitable to be included in this scheme in conjunction with the requirements of clause 7.2 of ISO 22000.

Thus there are no very special additional features in FSSC 22000, other than improving the prerequisite requirements and streamlining the structural issues. However, it further added the new guidelines and compliance criteria which increase the compliance levels above the minimum requirements which is one of the major differentiations of any private standard.

Thus, there is no point of discussing these private standards in detail, because it was already done under the ISO 22000 to give you an understanding of the structure which is almost equal for all the food safety standards practiced.

Good Manufacturing Practices

Food safety was primarily regulated since mid-1800s, but it was mostly the responsibility of the local or state regulations in US at the time [108]. The good manufacturing practices (GMP) were a result of requirement for consumer protection which is a set of regulations issued by authority of the Federal Food, Drug and Cosmetic Act. In the beginning of 20th century, there was no any federal regulation for the protection of public from dangerous products and most common preservation is cooling; thus, refrigeration means use of ice   where technology was primitive.   Actually, the situation was unthinkable where no pasteurization of milk, chemical   preservatives,   toxic   colours   and uncontrolled medicines on the market were contained with narcotics such as heroin, opium, and cocaine. In 1903, the Poison Squad was started raising on awareness on need for food safety lead by Harvey W. Wiley, a chemist working in USDA.

In 1906, Upton Sinclair published "The Jungle", a graphic exposure of the meat packing industry which was lead to pass "Pure Food and Drug Act of 1906". The 1906 law prevented interstate and foreign commerce in misbranded or adulterated foods, drinks, or drugs. The intent of the Act was to prevent poisoning and consumer fraud [109]. There were many tragedies occurred afterwards which lead to further strengthen and extend the regulations by passing The Federal Food, Drug and Cosmetic Act in 1938 [108]. There were drug amendments in 1962, where GMP regulation were grew due to many tragedies and accidental disasters throughout the human advancements in the last century to become the current regulations. The 1962 amendments formalized the good manufacturing practices initially targeting drug manufacturing industry. However, the GMP regulations for food processing facilities were finally proposed in 1968 and three broad categories of interrelated issues arose during the development of the GMPs [110]. The GMP regulations were finalized in April of 1969 and published as Part 128 of the Code of Federal Regulations (CFR). In 1977, Part 128 was recodified and published as Part 110 of the CFR [111].

According to the General Principles of Food Hygiene of Codex Alimentarius [112], International food trades, and foreign travel, are increasing, bringing important social and economic benefits. But this also makes the spread of disease around the world easier. Considering the current global food market, food manufacturers have to decide which quality assurance system is most suitable to their specific situation and how this system should be implemented, because there are number of quality assurance systems are available such as GMP, HACCP, ISO 9001, ISO 22000, FSSC 22000 and the international technical standard of British Retail Consortium (BRC). These systems and their combinations are applied for assuring food quality [113]. On the other hand, GMP consists of fundamental principles, procedures and means needed to design a suitable environment for the production of food of acceptable quality [114]. The basic aim of the GMP codes is to combine procedures for manufacturing and quality control in such a way that products are manufactured consistently to a quality appropriate to their intended use [115]. Nevertheless, regulatory requirements for a well-designed GMP program is varied by the type of product being produced and by the position of the product in the manufacturing process and supply chain where it is important for all food manufacturers to understand the appropriate GMP for their individual products [116].

A properly designed GMP system must have an appropriate infrastructure or "quality system", encompassing the organizational structure, procedures, processes and resources. It also needs to have systematic actions necessary to ensure adequate confidence that a product (or service) will satisfy given requirements for quality [117]. Based on this principle, GMPs are being applied to maintain the certainty of safety in the final product where it claims for the minimum sanitary and processing requirements to ensure the production of wholesome food. These requirements can be categorized in to four segments which are production and process controls, personnel practices, building facilities, equipment and utensils. On the other hand, all the employees involved in food supply chain to manufacture, pack and distribute food products for the consumption of humans, has to understand basic principles of sanitation with skills for application. Thus, human factor is one of most important criteria in GMP, where employee hygiene is paramount to plant sanitation and is one of the leading causes of food contamination [118]. In addition, manufacturers are legally liable to take necessary measures and precautions for disease control, personnel cleanliness, supervision, education and training to comply with GMP.

Nevertheless, every country has established minimum requirements for food manufacturing facilities which are almost in all occasions fully compatible with General Principles of Food Hygiene of the Codex Alimentarius or it was converted to national standards. Thus, General Principles of Food Hygiene is considered as the most common and comprehensive GMP document and applied to food manufacturing organizations across the globe, whereas the Section 32, Food Act, No. 26 of 1980 Sri Lanka is also having the same features. The main sections of the general principles of food hygiene has considered for;

I. Organization and Management Responsibilities,
II. Establishment, Design and Facilities,
III. Storage Facilities,
IV. Distribution Facilities,
V. Cleaning,
VI. Pest Control System,
VII. Personal Hygiene,
VIII. Quality Assurance System,

Based on the appropriate hygiene management of above subsectors of the production process, management has to perform periodical internal as well as third party audits or inspections to validate the application of GMP [119]. Record keeping is a necessity for GMP to monitor the compliance of the system. In addition, safety of the final product has liability to every personnel taking place in every step of processing line, who is responsible for the production of safe food which has to be well explained to the employee. Nonetheless, training of the personnel in every step of the line is important and the management must have goal for their product, which must start from the purchase of the material and continue through processing and distribution. This goal must be well understood by the every single personnel of the establishment, because GMP is a continuous process and any negligence in one of the steps will result with an inadequate and unsafe food product.

The 5S

The 5S is a standardized process, once properly implemented can creates and maintains an organized, safe, clean and efficient workplace, where improved visual controls are implemented as part of 5S to make any process non-conformances obvious and easily detectable. Nonetheless, 5S is also use as a major element of Lean initiative which promotes continuous improvement. The 5S Principles are recognized in many industries as effective tools where tea industry is no exclusion for improving workplace organization, reducing waste and increasing efficiency of manufacturing process. However, 5S should be use in collaboration with other management practices to yield better results, otherwise it could become an end goal of the company's improvement process instead of a key part of a larger continuous improvement journey.

The currently many tea manufacturers in the country are practicing 5S, because it was a part of the mandatory requirements of achieving SLTB's star classification few years back which died afterwards. However, many factory owners embrace it and kept continuing as a common practice, non like other standards in tea industry, but what they haven't realized was the greatest benefit from using 5S is realized when it is part of a larger initiative and the entire organization has adopted its principles. 5S is more than just a system, which is a business philosophy and should be integrated into the organization's culture to yield better results. The 5S will be discussed deeply in the latter stages of the book in related to local tea industry.

The 5S list is as follows:
Seiri / Sort – Separating of the essential from the nonessential items
Seiton / Straighten – Organizing the essential materials where everything has its place
Seiso / Shine – Cleaning the work area
Seiketsu / Standardize – Establishing a system to maintain and make 5S a habit

Shitsuke / Sustain – Establishing a safe and sanitary work environment (Safety)

Organic Certifications

The organic agriculture is a sustainable form of agriculture, for both ecological and economic reasons which produces products with the traditional methods while preserving the environment and avoiding most synthetic materials such as artificial fertilizers, pesticides and antibiotics. The organic certifications are based on the organic guidelines, regulations and standards developed by the International Federation of Organic Agriculture Movements (IFOAM) and Codex Alimentarius. However, there are slight variations on organic certifications from one country to another, but their core requirements are similar. Considering the organic agriculture, organic production is an overall system of farm management and food production which combines;

Best environmental practices;

High level of biodiversity;

Preservation of natural resources;

Application of high animal welfare standards;

Production method in line with the preference of certain consumers for products produced using natural substances and processes;

The certification process ensures that it preserve natural resources and biodiversity, support animal health and welfare, provide access to the outdoors so that animals can exercise their natural behaviors, only use approved materials, do not use genetically modified ingredients, separate organic food from non-organic food and receive annual onsite inspections, where organic production method plays a dual societal role. Basically organic agriculture caters for a specific market responding to a consumer demand on organic products while delivering public goods contributing to the protection of the environment and animal welfare as well as rural development.

The organic certification is a certification process where organic food and other organic agricultural products are certified according to the certain guidelines laid by the importing country. But the specific programs are having common agreements which includes avoidance of synthetic chemical inputs such as fertilizer, pesticide, antibiotics, food additives and irradiation or the use of sewage sludge. It is also mandatory to use farmlands that are fee of prohibited chemical inputs for a number of year (often three years or more) and avoid genetically modified seeds. In the case of live stocks, there are specific guidelines to follow for feed, housing and breeding. It is also mandatory to keep all the records on production and sales for the verification proposes as well as maintaining strict physical separation of organic products from non-certified products with undergoing periodic on-site inspections. In some countries, the certification regulated by the government where commercial use of the term organic is legally restricted whereas organic producers are also subjected to the same agricultural, food safety, quality and other specific regulations which applies to non-organic producers.

Future of Food Safety Standards

Considering the current trends in the industry, as GFSI considered the PRP issue which was highlighted in many forums but unfortunately ISO didn't address the issue where GFSI got the millage and further strengthen their standard (FSSC 22000). Now they have added ISO 17025:2005 to the list of guidelines where the company does not need to accredit their internal laboratory, but they need to follow the guidelines to comply with it. The advantages are better control of the food safety and reliability of the certification to the end user. In contrast, there are significant variations in food safety regulations across the globe and among value chains which increase the burden of auditing costs of certifications on food manufacturers, as retailers require different certification frameworks to qualify suppliers. The impacts of these variations on relevant actors present practical reasons for the need of harmonizing food safety regulations [120] which are justifiable reasons that explains these variations. Some of these reasons are attributed to the distinct tastes, diets, or income levels and perceptions that influence the tolerance of populations towards the risk associated with food.

Alternatively, this will tend to increase the product price and the accumulated cost for the production than it disserves where companies may tend to let down these practices while make sure auditor meet the minimum requirement. Thus, consumer safety is paramount when it comes to food safety regulation; yet, regulators required to conduct due assessments of food safety risks on consumers as well as cost implications of enforcement strategies on industry to help mitigate costs incurred by industry, without compromising consumer safety [121].

Considering the behaviour of enterprises, whether enterprises respond to standards in a positive or negative manner depends on a variety of factors e.g. sector, enterprise size, financial situation and level of risk adversity, which suggest that the response of enterprises is not automatic and it reflects the interplay among different types of incentives operating at the level of mandated government regulation, pressure from the markets and liability laws [122], [123], [124]. Therefore, addition of different extra guidelines will be good as well as bad, because most of the current food safety certifications available in developing countries do not fully complying with any of the available food safety regulations, this may be the case for even developed countries, where audit firms are also managing a business and they very rarely suspend any system they accredit. Even in the process of accreditation, major food safety issues and critical food safety violations are mostly recorded as minor food safety violations.

In addition, most of the certification firms at the beginning (startup face in any country office) lose their controls to gain the market and to attract more customers because all these system certifications have become marketing tools. The food safety issues are further intensified due to such situations as well as auditing practices. The auditor verifies only a fraction of the system while most of the companies only comply with food safety requirements on the day of certification or surveillance audits. As previously discussed, different private standards introduced by brand manufacturers (i.e. FSSC 22000, BRC) and retailers will further introduce more variations into existing food safety regulations and the modes of conformity assessments [125] to improve and differentiate their standards for one another.

Currently, FSSC 22000 has introduced unannounced audits to their audit schemes which will be absorbed into other certification systems very soon. The initiative has very valid point as it definitely effect the health of the company. But it will further escalate cost of production as well as burden on the manufacturers to leave such certifications in the failure of adaptation. Hence, private standards will have limitations in their activity where certification bodies will most probably will conduct an unannounced audit to comply the certification, but hopefully they will probably will let their clients know the range at least rather than the exact audit date, since they want to keep the business running.

Thus a common reference point is required, explaining from where the process of harmonization of standards could be started, to reduce multiple certifications on food enterprises. As an initiative, SPS agreement which was introduced by the WTO facilitates a move towards this much needed common reference point, by providing a basis to establish equivalence and harmonize food safety regulations [121]. However, there is a long way to go ahead, because all the certification systems are profit oriented whether their initiation organization is labeled as non for profit or for profit.

Social Standards
As already discussed, various social accountability standards are operating in tea trade such as SA 8000, Ethical Tea Partnership, Rainforest Alliance, UTZ and Fair Trade etc. These standards are basically considered on environment, labour regulations, biodiversity and wellbeing of the producers as well as sustainability of the supply chain. All these standards are voluntary in nature and operated

through non-for-profit organizations. However, these are based on developed countries where there is an additional weightage on the buying decision due to the niche consumers in developed markets. The buyer decisions are usually depends on market behaviours, where European, Japanese and North American buyers are sensitive to these issues and they are ready to pay premium prices on the right products but manufacturers have to obtain the relevant certifications.

ETP

The Ethical Tea Partnership (ETP) is a not-for-profit organization which was initiated in 1997 and originally called the Tea Sourcing Partnership. The initiative was established by number of large UK based tea packing companies who decided to work together with producers to improve tea sustainability, the lives and livelihoods of tea workers and smallholder farmers, and the environment in which tea is produced. According to the vision of the organization, it says "our vision is of a thriving tea industry that is socially just and environmentally sustainable". The organization has 40 international member companies, global tea brands, boutique labels and retailers with an international expert staff in the major tea manufacturing regions of the world engaging with more than 1000 estates/producers while supporting 700,000 farmers around the world. The difference in this certification is that it is free for producers.

The core areas of operations are;

1. Raising core standards

   The improvements of standard ensures producers in supply chains meet international social and environmental standards, where members must improve conditions for workers, and the way tea estates and factories are managed.

2. Improving worker lives and their livelihoods

   The organization conducts training and support programs which make work places better, fairer, and safer and ETP usually work with partners on poverty reduction and work towards the Millennium Development Goals. ETP also conduct programs which identify and mitigate complex time sensitive issues that go beyond the boundaries of the tea sector.

3. Improving Smallholder Lives and Livelihoods

   ETP help smallholders to achieve better incomes through improved quality and productivity, access to international markets, and affordable finance.

4. Climate Change and the Environment

   ETP further help producers improve their environmental management systems to protect soil, water, ecosystems, and wildlife where producers and smallholders understand the implications of climate change and ways to maintain production in the face of changing weather patterns and growing conditions.

The ETP global standard complies with all the key social compliances including all the relevant International Labour Organization requirements, environmental compliances and help producers to comply with all relevant recognized standards. The ETP operates through London based office with regional managers in China, India, Indonesia, Kenya and Sri Lanka. The following core areas under their supervision, which are closely aligned to the other main certification programs operating in tea such as Fairtrade, Rainforest Alliance and UTZ Certified. Additionally, ETP works to minimize the duplication of programs with above programs while creating harmony and support to implement following areas as core requirements in social and environmental disciplines.

Social Requirements
- Freely chosen employment
- Freedom of Association and the Right to Collective Bargaining
- Health and Safety
- Child Labour
- Wages and Benefits
- Working Hours
- Discrimination
- Regular Employment
- Disciplinary Procedures

Environmental Requirements
- ☯ Environmental Management Systems
- ☯ Agrochemicals
- ☯ Soil Conservation
- ☯ Ecosystem Conservation
- ☯ Water Conservation
- ☯ Energy Use
- ☯ Waste Management

Fairtrade Certification

Fair trade is a social movement with an alternative approach to conventional trade which is based on the partnership between producers and the consumers whose primary goal to help producers in developing countries to achieve better trading conditions and to promote sustainability. Fairtrade offers consumers a powerful way to reduce poverty through their daily shopping where movement advocate the payment of premium prices to exporters as well as improved social and environmental standards. The Fairtrade has focused on particular commodities or products which are typically exported from developing countries to developed countries such as handicrafts, coffee, cocoa, tea, wine, fresh fruit, chocolate, flowers, gold, etc. where Fairtrade strive to promote greater equity, in international trading partnerships through dialogue, transparency and respect. The movement promotes sustainable developments through offer of better trading conditions with secured rights for marginalized producers and workers in developing countries.

The Fairtrade standards are designed to address the imbalance of power in trading relationships, unstable markets and the injustices of conventional trade where it is grounded in three core beliefs:

> Producers have the power to express unity with consumers;
>
> The world current trade practices promote the unequal distribution of wealth between nations;
>
> Purchasing products from producers in developing countries at a fair price is a more efficient way of promoting sustainable development than traditional charity and aid.

The Fairtrade movement first began when several religious and humanitarian organizations got together with a common goal of helping impoverished workers in some of the world's poorest regions where their idea was to help improve wages and living conditions for disadvantaged workers. Further, movement wanted to help rebuilding the lives of displaced refugees, as well as improve conditions following natural disasters. The United Nations Conference on Trade and Development announce "trade not aid" in 1968 at New Delhi which set the groundwork for the phenomenal growth of Fairtrade and five years later, in 1973 the first fairly traded coffee was imported by Fairtrade Organization in the Netherlands, from cooperatives of small farmers in Guatemala. The market grew afterwards rapidly and today the bulk of merchandise sold consists of Fairtrade food products in the United Kingdom and it is over 30% in the U.S. imported food sales, which traditionally constituted less than 10%.

Fairtrade certification guarantees the principles of ethical purchasing and fair prices to the producer, while adhering to ILO agreements such as banning child labour, slave labour, safe work place with rights to unionize, adhere to charter of human rights of the United Nations, a fair price covering cost of production while facilitating social development, protection and conservation of the environment. The Fairtrade certification system further attempts to create sustainable long-term trade relationships between buyers and sellers as well as promotes crop pre-financing and proper transparency throughout the supply chain. The companies who offer fair products which comply with Fairtrade standards can apply for license to use on of the Fairtrade certification marks on the packaging where a product carries the Fairtrade Mark on it, that means the producers and traders have met Fairtrade Standards.

The Fairtrade operates in a fairer way, where packers in developed countries pay a subscription fee to use the brand and logo where fee is almost all spent on marketing. The packers and retailers have the right to charge their consumers as to their costs involved, but the product must come from a certified Fairtrade cooperative and there is a minimum price at the time of market over saturated. In addition, the buyers pay additional 10c per each lb of the product purchased for community development projects, but cooperative has an upper ceiling of only one third of their produce and the rest must be sold at market prices. The exporting cooperative has right to use money in several ways, such as coverage of conformity costs and certification, other additional costs incurred or annual turnover. Rest of the money or whole amount can also invest in community development projects.

Occupational Health and Safety
Occupational safety and health (OSH), also commonly known as occupational health and safety (OHS), occupational health or workplace health and safety (WHS), is a multidisciplinary filed concerned with safety, health and welfare of people at work. As to the further explanations given in Wikipedia; in common-law jurisdictions, employers have a common law duty to take reasonable care of the safety of his employees. Statue law may in addition impose other general duties, introduce specific duties and create government bodies with vested powers to regulate workplace safety issues where

details of this vary from jurisdiction to jurisdiction. In common sense, all organizations have the duty to ensure that employees and any other person who may be affected by organization's activities remain safe all times.

ISO 14001

The ISO 14000 is a family of standards which provides practical tools for companies and organizations of all kinds looking to manage their environmental responsibilities, where ISO 14001 is the most popular out of the family and known to the general public as an EMS. Hence ISO 14001:2015, the most recent revision of the standard is applicable to any organization, regardless of size, type or nature, which applies to the environmental aspects of its activities, products and services that the organization determines, that can either control or influence considering a life cycle perspective. Nonetheless, ISO 14001 does not state specific environmental performance criteria, which can be used in whole or in part to systematically improve environmental management, but it is not acceptable unless all of its requirements are incorporated into the organization's environmental management system and fulfilled without exclusion.

The first ever published environmental management system standard was BS 7750 of BSI group in March, 1992 to protect the environment while responding to the growing industry concerns.

The development of BS 7750 lead to the development of ISO 14000 series in 1996 where voluntary models of the industry groups as well as initiatives such as Responsible care went out of industry due to difficulties in assessment, compliance and certification which included usual ISO standard requirements such as environmental policy, organizational structure, planning activities, responsibilities, practices, procedures, processes, resources for the development, implementation, maintenance and improvement of the EMS. However, the standard improved with time and changed its structure to the most current version of it by migrating to Annex SL format. Today, standard is accepted all over the world with more than 300,000 certifications in 171 countries worldwide according to the ISO survey results of 2017.

ISO 14001 is a generic standard which is a voluntary that provides the framework for organizations to demonstrate their commitment to the environment by reducing harmful effects on the environment, while meeting environmental legal requirements and providing evidence of continual improvement of environmental management. The standard is generic in nature and does not apply to any particular industry or business sector, which usually provides a strategic framework that can be used to meet internal and external objectives for environmental management. As all the ISO standards are agrees on same high level structure today, it usually follow the same standard requirements which can be implemented in a multiple platform with collaboration of other ISO standards.

ISO 14001 provide guidelines for any organization seeking to improve and manage resources more effectively, without any limitation from single-site to large multi-national companies, high-risk companies to low-risk service organizations. Nonetheless, it further apply to any manufacturing, process, and service industries, including local governments as well as all industry sectors, including public and private sectors and original equipment manufacturers and their suppliers. In addition, the current version of the ISO 14001:2015 is more focused on improving the environmental performance, instead of improving the management system when it is migrated to new structure.

Following are a list of ISO 14000 family of standards:

1. ISO 14004 – General guidelines on principles, systems and support techniques
2. ISO 14006 – Guidelines for incorporating eco-design
3. ISO 14015 – Environmental assessment of sites and organizations (EASO)
4. ISO 14020 – Environmental labels and declarations
5. ISO 14031 – Environmental performance evaluation
6. ISO 14040 – Life cycle assessment

7. ISO 14050 – Vocabulary
8. ISO 14063 – Environmental communication
9. ISO 14064 – Greenhouse gases
10. ISO 19011 – Guidelines for auditing management systems

6
Occupational Health and Safety in Tea Industry

Production of tea consists of various mechanical and manual operations throughout the processing which is very complicated after drying is done. Thus, grading is the most labour intensive part of the processing which has been automated up to a large extent, but it still needs various manual works to get good grades. Usually tea is packed in wooden crates in the past which has being changed to paper sacks today in most of the exporting countries, but it is still practiced. The made teas are further blended in consuming countries, where various different products are mixed together to create brand specific blends by the packer. Most of the small granules teas are send to the packing machines for single tea bags while others send for bulk packing machines or consumer packs of loose teas. Considering the entire process, most of the occupational hazards are occurs during the initial processing stages such as withering, rolling, fermentation, drying, grading, blending and packing. The common health and safety hazards are occurring due to improper machine grading, noise, dust, slips, falls, and lifting-related injuries.

Machine/Equipment Hazards

The machine related health and safety hazards are occurring due to worker's involvement and exposure to chain and sprockets, belts and pulleys, rotating shafts and equipment, roll breakers, dryers as well as high-speed packing lines containing number of risky or dangerous pinch points. The most common injuries are the results of lacerations or bruises to hand, fingers and arms where grading machinery and equipment are the critical control points to safeguard operators which minimize getting caught in, under or between moving parts. Thus, guards and/or interlocks must be installed based on the health and safety risk assessments carried out on the production line. The maintenance or cleaning is one of the critical operations where all energy sources have to be isolated while removing operators from production floor other than technicians and there should be an effective lockout during maintenance or cleaning operations. The operation has to be properly tag-out with visual aids to avoid accidents during this time.

Noise Hazards

Most of the machines designed for tea industry are century old and very few modifications and automations had been done to increase productivity. Additionally, high-speed packing lines usually generate high noise levels too. Generally, these machines generate high noise levels over stipulated industry norms, which is harmful to human ear and it pollute surrounding environment. Furthermore, the high noise levels are usually generated from vibrating or moving parts, rollers, withering fans, air conveying systems, dust collectors, exhaust fans, packing machines as well as box cutters. The noise levels are usually range from 85 dBA to over 90 dBA and usually affect due to the long-term exposure which typically a major potential critical health hazard lead to permanent hearing loss. The severity of the hazard is generally the noise level and the individual's susceptibility, while occurrence is exposure time, but when you need to access the criticality exposure time and the noise level is considered.

Material Handling
Material handling is starting from the moment tea leaves are detached from the tea bush, thus they are usually packed in plastic crates, gunny/poly sacks, from the plantations and carried over by men or women if it is not automated. Even after automated lifting to the loft areas, tea leaves have to be packed into withering troughs where manual lifting is occupied. Thus, rest of the operations has many lifting in each and every processing operation. The lifting is further involved in blending and packing as well as loading, unloading through manual or mechanical means, where injuries to worker's lower back is very common when handling made tea weighing around 100 pounds (45.5kg) or more and repetitive motions on crates in grading rooms,

packing line or leaf bags can result in cumulative trauma to the wrist, arm and/or shoulder area. Some of this kind of lifting tasks can be altered with mechanical lifting devices such as vacuum lifts or trolleys. Furthermore, use of two workers, instead of single worker in repetitive heavy lifting operations will reduce the impact while reducing the cumulative effects on the serious back injuries to the workers or shifting work stations continuously within heavy lifting to light duties and back to repetitive tasks will also can reduce the risks. Most suitable options are to modify the work stations to be more ergonomically correct and fully automating them will reduce the heavy lifting throughout the processing operations. Provision of use of personal aids such as back-belts and wrist bands for lifting operators may assist in lifting tasks or for temporary relief of minor stains which has not been proven to be effective and they may be even harmful.

On the other hand, miss-use of fork-lift trucks or failure to drive at safe speeds, sharp turns, driving with raised forks, failure to observe or yield to pedestrians and loading/unloading accidents are the leading causes of injuries involving fork-lift operators where management must permit only use of trained and competent operators to drive fork-lifts. Operator training must consist of formal classroom training as well as formal driving tests where operators can demonstrate their skills. Nonetheless, proper maintenance and daily pre-use inspections also help to ensure the safe operation of such vehicles.

Exposure to High Temperatures

The tea manufacturing process generally required high energy sources due to the requirements of elevated temperatures for drying of tea at around 110°C where various methods are used to provide heat to the ovens. Thus, skin contact of firewood, steam lines, hot water as well as process equipment can result serious injury from burns which usually occurs to the hands, arms and face. Use of hot water for cleaning or wash-down has also known to be causing burns on feet and legs. Furthermore, heat sealers as well as gluing of packing lines are also can cause burns.

Prevention of such hazards are very important where guarding of exposed hot surfaces are mandatory. Proper hazard analysis is to be conducted beforehand and selection and use of personal protective equipment, will help reduce or eliminate worker exposure to high temperatures and accidental burns. In addition, use of pipeline breaking and lockout procedures will protect workers from the unexpected release of hot liquids and steam.

Slips, Trips and Falls

Slips, trips and falls are a major concern where water, dust, oil or grease spills dry blending or packaging operations as well as fine tea dust will accumulate on walking and working surfaces. The prevention plans require good housekeeping and floors should be swept clean of tea dust on a regular basis. The "clean as you go" practices or in other words, debris collection has to be more efficient while slip-resistant, rubber-soled shoes appear to provide the best traction. Wet-process areas are extremely susceptible for slip and fall hazards where floors should be kept as dry as possible while improving adequate floor drainage within all wet-process areas. Nonetheless, water should not be permitted to accumulate where standing water exists, it should be mopped into floor drains.

Chemical Hazards

Tea manufacturing has no chemical usage as a raw material or ingredient in the production processes and packaging operations where workers have no threat of exposing to hazardous chemicals, but sanitation operations use chemicals to clean and sanitize equipment and floors. Some of the cleaning chemicals such as caustic soda or other detergents are handled in bulk quantities through fixed pipe systems or applied by hand using predetermined mixtures. Exposure to such chemicals can cause respiratory problems, dermatitis or skin irritation and chemical can produce burns to the skin where severe burns to the eyes and/or loss of vision are the critical hazards associated with cleaning chemicals. The initial hazard analysis on the cleaning chemicals and proper evaluations are essential to mitigate chemical hazards. As a prevention strategy, proper selection and use of personal protection equipment (PPE) should be part of routine job procedure where PPE such as splash-proof goggles or face shields, chemical-resistant gloves, aprons, boots and a respirator can be applicable based on the application and the safety requirements given with the respective chemical. Nonetheless, emergency eye and body wash stations should be provided where hazardous chemicals are either stored, mixed or used.

Dust Hazards

Dust and micro-fibers are one of the major outcomes of tea manufacturing process where significant amounts of tea dust and micro-fibers are generated starting from drying which continue through grading, blending and packaging operations. Additionally, tea dust may also be present in high concentrations during clean-up or blow-down operations as well as due to the breakdowns. In fact, tea dust with a diameter greater than 10 micrometers can be classified as "nuisance dust" where nuisance dust has little adverse effect on the lungs and should not produce significant organic disease or toxic effects when exposures are kept under reasonable control. However, excessive concentrations of nuisance dust in the workroom air, which may cause unpleasant deposits in the eyes, ears and nasal passages. Furthermore, once these particles are inhaled that may become entrapped in the nasal and pharyngeal region of the respiratory system, until they are expelled through the body's own cleaning mechanisms (e.g., coughing or sneezing).

Hence respirable dust particulates are those that are less than 10 micro-meters in diameter which is small enough to pass through the nasal and pharyngeal regions to enter the lower respiratory tract. Once they reach the lungs, dust may become embedded in the alveolar region, where scar tissue may perhaps develop. Thus respirable particulates can be respiratory irritants, especially in asthmatics where effective seals and closures will help contain dust particles.

Prevention of dust inhalation can be done with exhaust ventilation or other types of dust-control equipment which can be provided at the site of dust generation to maintain dust levels below the internationally reputed standards (10 mg/m3) or other country specific government regulations in force. The workers shall be provided with dust masks and it must be worn by workers who may be highly sensitive to dusts and workers who exposed to higher concentrations of dust at any one time. Furthermore, vulnerable people such as persons with chronic bronchitis or asthma are at higher risk where they need to be keep away from such production areas as well as workers who suffer from hypersensitivity to tea dust should be removed from the area.

Next possible hazard is tea dust explosion although there is little information available on actual tea dust explosions, as to the published test data, the explosion characteristics of tea dust are relatively weak where greatest potential for a tea dust explosion exists with storage bins and dust collectors due to optimization of concentrations and particle size. The prevention strategies must include minimizing dust concentration within a room or process will reduce the potential of a dust explosion, whereas electrical equipment designed for dust hazard areas may also be desirable in some operations. There are minor possibilities of tea and tea dust burst into flames, where large quantities of tea will almost always smoulder if ignited. Thus large quantities of water in a fine mist can be used to cool the smouldering tea below its ignition temperature in an emergency situation.

General Safety Practices
A general safety program must address the gaps identified during the evaluation process where use and selection of PPE, entry into confined spaces, isolation of energy sources, identification and communication of hazardous chemicals, self-inspection programs, hearing conservation programs, the control of infectious materials, process management, emergency preparedness and

response programs must also be included as part of the work process. Training of workers in safe work practices is important in reducing worker exposure to hazardous conditions and injuries where annual training and re-training events are mandatory to update the worker awareness on health and safety issue in the tea trade.

7

Total Quality Management

Total quality management (TQM) consists of organization-wide efforts to install and create permanent climate in which an organization continuously improves its ability to deliver high quality products and services to customers. While there is no widely agreed upon approach, TQM efforts typically draw heavily on the previously developed tools and techniques of quality control. TQM enjoyed widespread attention during the late 1980s and early 1990s before being overshadowed by ISO 9001, Lean manufacturing, and Six Sigma.

In the late 1970s and early 1980s, the developed countries of North America and Western Europe suffered economically in the face of stiff competition from Japan's ability to produce high quality goods at competitive cost. For the first time since the start of the Industrial Revolution, the United Kingdom became a net importer of finished goods. The United States undertook its own soul searching, expressed most pointedly in the television broadcast of "If Japan Can... Why Can't We?" firms began reexamining the techniques of quality control invented over the past 50 years and how those techniques had been so successfully employed by the Japanese. It was in the midst of this economic turmoil that TQM took root.

The Quality

The definition of quality depends on the role of the people defining it. Most consumers have a difficult time defining quality, but they know it when they see it. For example, although you probably have an opinion as to which manufacturer of athletic shoes provides the highest quality; it would probably be difficult for you to define your quality standard in precise terms. Also, your friends may have different opinions regarding which French Fries are of highest quality. The difficulty in defining quality exists regardless of product, and this is true for both manufacturing and service organizations. Think about how difficult it may be to define quality for products such as airline services, child day-care facilities, college classes, or even textbooks. Further complicating issue is that, the meaning of quality has changed over time. Today, there is no single universal definition of quality. Some people view quality as "performance to standards." Others view it as "meeting the customer's needs" or "satisfying the customer." Let's look at some of the more common definitions of quality.

Conformance to specifications measures how well the product or service meets the targets and tolerances determined by its designers. For example, the dimensions of a machine part may be specified by its design engineers as 3 ±.05 inches. This would mean that the target dimension is 3 inches but the dimensions can vary between 2.95 and 3.05 inches.

Similarly, the wait for hotel room service may be specified as 20 minutes, but there may be an acceptable delay of an additional 10 minutes. Also, consider the amount of light delivered by a 60 watt light bulb. If the bulb delivers 50 watts it does not conform to specifications and how about the 300ml of a Coke bottle which has ±5ml margin for accuracy. As these examples illustrate, conformance to specification is directly measurable, though it may not be directly related to the consumer's idea of quality.

Fitness for use focuses on how well the product performs its intended function or use. For example, a Mercedes Benz and a Jeep Cherokee both meet a fitness for use definition if one considers transportation as the intended function. However, if the definition becomes more specific and assumes that the intended use is for transportation on mountain roads and carrying fishing gear, the Jeep Cherokee has a greater fitness for use. You can also see that fitness for use is as a user based definition in that it is intended to meet the needs of a specific user group.

Value for price paid is a definition of quality that consumers often use for product or service usefulness. This is the only definition that combines economics with consumer criteria; it assumes that the definition of quality is price sensitive. For example, suppose that you wish to sign up for a personal finance seminar and discover that the same class is being taught at two different colleges at significantly different tuition rates. If you take the less expensive seminar, you will feel that you have received greater value for the price.

Support services provided are often how the quality of a product or service is judged. Quality does not apply only to the product or service itself; it also applies to the people, processes, and organizational environment associated with it. For example, the quality of a university is judged not only by the quality of staff and course offerings, but also by the efficiency and accuracy of processing paperwork.

Psychological criteria are a subjective definition that focuses on the judgmental evaluation of what constitutes product or service quality. Different factors contribute to the evaluation, such as the atmosphere of the environment or the perceived prestige of the product. For example, a hospital patient may receive average health care, but a very friendly staff may leave the impression of high quality. Similarly, we commonly associate certain products with excellence because of their reputation; Rolex watches and Mercedes-Benz automobiles are examples.

The term used for today's new concept of quality is total quality management or TQM. Consider the imagination of the old and new concepts of quality.

Indeed the difference is that the old concept is reactive, designed to correct quality problems after they occur. The new concept is proactive, designed to build quality into the product and process design.

The philosophy

What characterizes TQM is the focus on identifying root causes of quality problems and correcting them at the source, as opposed to inspecting the product after it has been made. Not only does TQM encompass the entire organization, but it stresses that quality is customer driven. TQM attempts to embed quality in every aspect of the organization. It is concerned with technical aspects of quality as well as the involvement of people in quality, such as customers, company employees, and suppliers. Let's look at the specific concepts that make up the philosophy of TQM.

Customer Focus

The first, and overriding, feature of TQM is the company's focus on its customers. Quality is defined as meeting or exceeding customer expectations. The goal is to first identify and then meet customer needs. TQM recognizes that a perfectly produced product has little value if it is not what the customer wants. Therefore, we can say that quality is customer driven. However, it is not always easy to determine what the customer wants, because tastes and preferences change. Also, customer expectations often vary from one customer to the next. For example, in the auto industry trends change relatively quickly, from small cars to sports utility vehicles and back to small cars. The same is true in the retail industry, where styles and fashion are short lived. Companies need to continually gather information by means of focus groups, market surveys, and customer interviews in order to stay in tune with what customers want. They must always remember that they would not be in business if it were not for their customers.

Continuous Improvement

Another concept of the TQM philosophy is the focus on continuous improvement. Traditional systems operated on the assumption that once a company achieved a certain level of quality, it was successful and needed no further improvements. We tend to think of improvement in terms of plateaus that are to be achieved, such as passing a certification test or reducing the number of defects to a certain level. Traditionally, change for American managers involves large magnitudes, such as major organizational restructuring. The Japanese, on the other hand, believe that the best and most lasting changes come from gradual improvements. To use an analogy, they believe that it is better to take frequent small doses of medicine than to take one large dose. Continuous improvement, called kaizen by the Japanese, requires that the company continually strive to be better through learning and problem solving. Because we can never achieve perfection, we must always evaluate our performance and take measures to improve it.

Quality starts with market research to establish the true requirements for the product or service and the true needs of the customers. However, for an organization to be really effective, quality must span all functions, all people, all departments and all activities and be a common language for improvement. The cooperation of everyone at every interface is necessary to achieve a total quality of an organization, in the same way that the Japanese achieve this with companywide quality control.

Customers and Suppliers
There exists in each department, each office, every home, a series of customers, suppliers and customer supplier an interface. These are "the quality chains", and they can be broken at any point by one person or one piece of equipment not meeting the requirements of the customer, internal or external. The failure usually finds its way to the interface between the organization and its external customer, or in the worst case, actually to the external customer.

Failure to meet the requirements in any part of a quality chain has a way of multiplying, and failure in one part of the system creates problems elsewhere, leading to yet more failure and problems, and so the situation is exacerbated. The ability to meet customers' (external and internal) requirements is vital.

To achieve quality throughout an organization, every person in the quality chain must be trained to ask the following questions about every customer-supplier interface:

Customers (internal and external)
1. Who are my customers?
2. What are their true needs and expectations?
3. How do, or can, I find out what these are?
4. How can I measure my ability to meet their needs and expectations?
5. Do I have the capability to meet their needs and expectations? (If not, what must I do to improve this capability?)
6. Do I continually meet their needs and expectations?
7. (If not, what prevents this from happening when the capability exists?)
8. How do I monitor changes in their needs and expectations?

Suppliers (internal and external)
1. Who are my internal suppliers?
2. What are my true needs and expectations?
3. How do I communicate my needs and expectations to my suppliers?

4. Do my suppliers have the capability to measure and meet these needs and expectations?
5. How do I inform them of changes in my needs and expectations?

As well as being fully aware of customers' needs and expectations, each person must respect the needs and expectations of their suppliers. The ideal situation is an open partnership style relationship, where both parties share and benefit.

Poor Practices
To be able to become a total quality organization, some of the bad practices must be recognized and corrected. These may include:

1. Leaders do not give clear direction
2. Do not understand or ignore competitive positioning
3. Each department working only for itself
4. Trying to control people through systems
5. Confusing quality with grade
6. Accepting that a level of defects or errors is inevitable
7. Firefighting, reactive behaviour
8. The "It's not my problem" attitude

How many of these behaviours do you recognize in your organization?

Use of Quality Tools

Continuous quality improvement process assumes and even demands that a team of experts in field as well as a company leadership actively use quality tools in their improvement activities and decision making process. Quality tools can be used in all phases of production process, from the beginning of product development up to product marketing and customer support. At the moment, there are a significant number of quality assurance and quality management tools on disposal to quality experts and managers, so the selection of most appropriated one is not always an easy task. In the conducted research it is investigated possibilities of successful application of 7QC tools in several companies in power and process industry as well as government, tourism and health services. The seven quality tools are:

1. Cause and Effect Diagrams
2. Flow Charts
3. Checklists
4. Control Charts
5. Scatter Diagrams
6. Pareto Analysis
7. Histograms

You can see that TQM places a great deal of responsibility on all workers. If employees are to identify and correct quality problems, they need proper training. They need to understand how to assess quality by using a variety of quality control tools, how to interpret findings, and how to correct problems. In this section we look at seven different quality tools. These are often called the seven tools of quality control and they are easy to understand, yet extremely useful in identifying and analyzing quality problems. Sometimes workers use only one tool at a time, but often a combination of tools is most helpful.

Cause and Effect Diagrams
Cause and effect diagrams are charts that identify potential causes for particular quality problems. They are often called fishbone diagrams because they look like the bones of a fish. The "head" of the fish is the quality problem, such as damaged zippers on a garment or broken valves on a tire. The diagram is drawn so that the "spine" of the fish connects the "head" to the possible cause of the problem. These causes could be related to the machines, workers, measurement, suppliers, materials, and many other aspects of the production process. Each of these possible causes can then have smaller "bones" that address specific issues that relate to each cause. For example, a problem with machines could be due to a need for adjustment, old equipment or tooling problems. Similarly, a problem with workers could be related to lack of training, poor supervision, or fatigue. Cause and effect diagrams are problem solving tools commonly used by quality control teams. Specific causes of problems can be explored through brainstorming. The development of a cause

and effect diagram requires the team to think through all the possible causes of poor quality.

Flow Charts

A flowchart is a schematic diagram of the sequence of steps involved in an operation or process. It provides a visual tool that is easy to use and understand. By seeing the steps involved in an operation or process, everyone develops a clear picture of how the operation works and where problems could arise.

Checklists

A checklist is a list of common defects and the number of observed occurrences of these defects. It is a simple yet effective fact finding tool that allows the worker to collect specific information regarding the defects observed. This means that the plant needs to focus on this specific problem; for example, by going to the source of supply or seeing whether the material is the issue; during a particular production process. A checklist can also be used to focus on other dimensions, such as location or time. For example, if a defect is being observed frequently, a checklist can be developed that measures the number of occurrences per shift, per machine, or per operator. In this fashion we can isolate the location of the particular defect and then focus on correcting the problem.

Control Charts

Control charts are a very important quality control tool. These charts are used to evaluate whether a process is operating within expectations relative to some measured value such as weight, width, or volume. For example, we could measure the weight of a sack of flour, the width of a tire, or the volume of a bottle of soft drink. When the production process is operating within expectations, we say that it is "in control." To evaluate whether or not a process is in control, we regularly measure the variable of interest and plot it on a control chart. The chart has a line down the center representing the average value of the variable we are measuring. Above and below the center line are two lines, called the upper control limit (UCL) and the lower control limit (LCL). As long as the observed values fall within the upper and lower control limits, the process is in control and there is no problem with quality. When a measured observation falls outside of these limits, there is a problem.

Scatter Diagrams

Scatter diagrams are graphs that show how two variables are related to one another. They are particularly useful in detecting the amount of correlation, or the degree of linear relationship, between two variables. For example, increased production speed and number of defects could be correlated positively; as production speed increases, so does the number of defects. Two variables could also be correlated negatively, so that an increase in one of the variables is associated with a decrease in the other. For example, increased worker training might be associated with a decrease in the number of defects observed.

The greater the degrees of correlation, the more linear are the observations in the scatter diagram. On the other hand, the more scattered the observations in the diagram, the less correlation exists between the variables. Of course, other types of relationships can also be observed on a scatter diagram, such as an inverted U. This may be the case when one is observing the relationship between two variables such as oven be the case when one is observing the relationship between two variables such as oven temperature and number of defects, since temperatures below and above the ideal could lead to defects.

Pareto Analysis

Pareto analysis is a technique used to identify quality problems based on their degree of importance. The logic behind Pareto analysis is that only a few quality problems are important, whereas many others are not critical. The technique was named after Vilfredo Pareto, a nineteenth century Italian economist who determined that only a small percentage of people controlled most of the wealth. This concept has often been called the 80 – 20 rule and has been extended into many areas. In quality management the logic behind Pareto's principle is that most quality problems are a result of only a few causes. The trick is to identify these causes.

One way to use Pareto analysis is to develop a chart that ranks the causes of poor quality in decreasing order based on the percentage of defects each has caused. For example, a tally can be made of the number of defects that result from different causes, such as operator error, defective parts, or inaccurate machine calibrations. Percentages of defects can be computed from the tally and placed in a chart. We generally tend to find that a few causes account for most of the defects.

Histograms
A histogram is a chart that shows the frequency distribution of observed values of a variable. We can see from the plot what type of distribution a particular variable displays, such as whether it has a normal distribution and whether the distribution is symmetrical.

In the food service industry the use of quality control tools is important in identifying quality problems. Grocery store chains must record and monitor the quality of incoming produce, such as tomatoes and lettuce. Quality tools can be used to evaluate the acceptability of product quality and to monitor product quality from individual suppliers. They can also be used to evaluate causes of quality problems, such as long transit time or poor refrigeration. Similarly, restaurants use quality control tools to evaluate and monitor the quality of delivered goods, such as meats, produce, or baked goods.

Quality Leadership
In many businesses competition on the market for customers is extremely intense. The situation could be likened to a war. The organizations that are successful in this struggle are those who are better and more efficient than others at satisfying customer needs and wishes. The war is not conducted solely on the domestic market, in some cases foreign markets are the most important arenas.

Success in battle requires good leadership. The same applies to acquiring customers who are entirely satisfied with the products. What is required is quality leadership. Leadership emanates from the senior manager in the organization. It is he or she who should be quality leader number one, and in that capacity provide hands-on leadership in the quality field. Leadership of this type is based on management's explicitly stated vision. This vision should describe what the company intends to achieve in terms of quality.

In order to realize this vision, real leadership is required, quality policy, quality goals and a quality system.

Because the quality policy describes the vision and provides brief guidelines for how the business is to be run to realize this vision. Thus, quality goals are specific, measurable goals for quality activities. It is particularly important to have goals for quality improvements. Quality work includes activities, procedures and methods. These form a network known as the quality system. Customers may demand that the company has a quality system in accordance with the requirements of ISO 9001 or ISO 22000 or both.

Quality of Organization
Quality is the outcome of the work for many people. If a good result is to be achieved, the organization should have a structure with a proper division of responsibility and authority for quality activities is clearly defined. All the people in the organization influence quality through their day-to-day work. It is therefore important that everyone is given conditions that will allow them to perform their work so that the right quality results. In many organizations shifting the focus onto quality represents a change in corporate culture. Characteristics of the new culture will include trust and delegation. All the employees become involved in a continuous improvement process.

Quality Policy

Effective quality leadership depends on management providing clear guidelines as to how the business is to be run and to follow up that the business is run in the stipulated way. These guidelines are best included in the company's written quality policy. According to the international standard for terminology, ISO 8402, quality policy is defined as "overall intensions and direction of an organization with regard to quality, as formally expressed by top management". The international standards for quality system requirements, which form part of the ISO 9001 state, as a first requirement, that a company should have a quality policy.

It is becoming increasingly common for companies to have a quality policy. The primary factor behind this is the growing interest in ISO 9001. If management wishes to act in accordance with the requirements of ISO 9001 this naturally means that the company must have a quality policy. If this is the only reason for drawing up a quality policy, the relevance of having one at all might be questioned. What is preferable is that management feels deeply that it wishes to express its intentions in the quality field to its employees. A quality policy should come from the heart.

Many organizations do not have a written quality policy. Even so, many of them believe that they do have a policy. What they have in mind in such cases is certain principles, which have emerged about the way things are done. These principles perhaps originate from decisions that top management has made on specific issues. Situations can arise where the absence of a clearly stated, generally known quality policy could obstruct activities intended to solve quality problems. The real reason why the organization finds itself in difficulties in the quality field could in fact be that it does not have a quality policy. If a quality policy is to have the desired effect on quality activities, certain points should be taken into account:

The quality policy should reflect the business idea, because underlying the business as such there is a business idea, a concept or a vision. The company may also have stated long-term objectives and guidelines for the business. The quality policy should naturally help to realize the ideas and objectives stated in these. This means that the quality policy should be in line with these objectives. Influencing the quality of a company's goods and services is a task, which should have a long-term focus. Hence, the quality policy should be long-term in character. More or less all functions have a direct influence on the quality of goods and services. Thus, quality policy should provide guidelines for the activities of all these functions, where quality policy should be comprehensive.

The contents of the quality policy should be of importance to the business. Unnecessary and vague formulations should be avoided, where quality policy should be relevant. In addition, if the quality policy is to achieve the desired effect it should be communicated to everyone in the organization. Hence, it should be written in simple, easy to-understand terms, where quality policy should be expressed in simple terms. If the quality policy is long, everyone won't read it, and it is much easier to read if it is short and to the point, where quality policy should be brief. Nonetheless, quality policy should be communicated in an authoritative manner. The quality policy should reflect the intentions of the company's management in the quality field and explain how in principle they are to be achieved. This means that the policy document should be issued by the chief executive. No other name, such as that of the quality manager, should appear on the document as this might give a signal that top management does not give its full support to the policy. For greatest effect, the quality policy should have major considerations.

1. **The need for a quality policy**
    This explains the background and objects of the quality policy.

2. **The quality level of the goods or services**
    Should the aim be to achieve leadership on

quality, leadership on price, or not even to be a leader at all, etc.?

3. **Customer relationships** – Analyses of customer needs, response to complaints.

4. **Supplier relationship** – Should suppliers be treated as an internal department and thus be given the equivalent support or should they be left to their own devices?

5. **Relationships with personnel** – Should the personnel be given all the conditions required to do a good job?

Risk Assessment

Risk assessment is the determination of quantitative or qualitative value of risk related to a concrete situation and a recognized threat (also called hazard) or dictionary definition for risk is possibility of suffering harm or loss; danger. Scientists use the term risk when assessing potential human health threats from exposure to chemicals or pollutants in the environment. Risk is equal to a person's exposure multiplied by the toxicity of the chemical. Quantitative risk assessment requires calculations of two components of risk (R), the magnitude of the potential loss (L), and the probability (p) that the loss will occur. Acceptable risk is a risk that is understood and tolerated usually because the cost or difficulty of implementing an effective countermeasure for the associated vulnerability exceeds the expectation of loss.

Risk assessment consists of an objective evaluation of risk in which assumptions and uncertainties are clearly considered and presented. Part of the difficulty in risk management is that measurement of both of the quantities in which risk assessment is concerned - potential loss and probability of occurrence - can be very difficult to measure. The chance of error in measuring these two concepts is large. Risk with a large potential loss and a low probability of occurring is often treated differently from one with a low potential loss and a high likelihood of occurring. In theory, both are of nearly equal priority, but in practice it can be very difficult to manage when faced with the scarcity of resources, especially time, in which to conduct the risk management process. The nuclear, aerospace, oil, rail and military industries have a long history of dealing with risk assessment. Also, medical, hospital, social service and food industries control risks and perform risk assessments on a continual basis. Methods for assessment of risk may differ between industries and whether it pertains to general financial decisions or environmental, ecological, or public health risk assessment.

In food industry, this methods are employed to calculate the risk in the identification of hazards and the critical control points for the system development in ISO 22000 or on the other words in HACCP system which is the food safety assurance system within the ISO 22000 structure.

8
Food Preservation

Food preservation in the broad sense of the term refers to all measures taken against any spoilage of food. In its narrower sense, however, food preservation connotes processes directed against food spoilage due to microbial or biochemical action. Preservation technologies are based mainly on the inactivation of microorganisms or on the delay or prevention of microbial growth. Consequently they must operate through those factors that most effectively influence the survival and growth of microorganisms.

Factors used for food preservation are called 'hurdles' and there are numerous hurdles that have been applied for food preservation. Potential hurdles for use in the preservation of foods can be divided into physical, physicochemical, microbiologically derived and miscellaneous hurdles. Among these hurdles, the most important have been used for centuries and are as either 'process' or 'additive' hurdles including high temperature, low temperature, water activity , acidity, redox potential (Eh), competitive microorganisms (e.g. lactic acid bacteria) and preservatives (e.g. nitrite, sorbate, sulphite). Recently, the underlying principles of these traditional methods have been defined and effective limits of these factors for microbial growth, survival, and death have been established. nonetheless, there are about 50 additional new hurdles have been used in food preservation in current context. These hurdles include: ultrahigh pressure, mano-thermo-sonication, photodynamic inactivation, modified atmosphere packaging of both non-respiring and respiring products, edible coatings, ethanol, Maillard reaction products and bacteriocins.

Barrier Technology
To control food safety, providing barriers to food contamination is a generally applied concept. The first barrier refers to outside premises, such as fencing, to prevent unauthorized access to the facility. The access of transport vehicles with raw materials and end-products, personnel, domestic and non-domestic animals should be monitored and controlled. Factory site drainage and storm water collection must be sufficient; areas within a 3m perimeter of the factory must be kept vegetation free to avoid pest breeding and harborage sites; a 10cm thick concrete curtain wall around the factory foundation at least 60 cm below ground discourages rodents from entering the building; effluent treatment plants and waste disposal units should be sited such that prevailing winds do not blow microbial and dust or aerosols into manufacturing areas.

The second barrier concerns the closing of factory buildings. All entrances/exits (i.e., window and door openings, openings for vents, air circulation lines, floor drains, etc.) must be designed for control over access, flow or exit of personnel, raw and finished food products, air, process aids (process water, process steam, food gases, etc.), waste, utilities (plant cooling and heating water, plant steam, compressed air, electricity, etc.) and pests (insects, birds, rodents, etc.). Floor drains must be screened to avoid rats from entering the food plant via sewers; ventilator openings, including vents in the roof, should be screened to prevent the entry of roof rats, insects and birds; gaps at the entrances of electrical conduits, process and utility piping, which are convenient pathways for roof rats, which must be closed.

The third barrier is the segregation of restricted areas (zones) within the plant, each of which has different hygienic requirements and controlled access. The fourth barrier is the processing equipment (including storage and conveying systems), which must have an adequate hygienic design and must be closed to protect the food product from external contamination. When the external contaminations have been eliminated, it is quit easier to handle internal contaminations as well as other sources of food spoiling while processing or after processing. Use of integrated preservation methods instead of using single preservation method to keep food fresh and nourished is a well-known fact, where it can further reduce the production cost as well as rigorous processing methods that's where Hurdle Technology come to play a major role.

Hurdle Technology

Hurdle technology was developed several years ago as a new concept for the production of safe, stable, nutritious, tasty and economical foods. It advocates the intelligent use of combinations of different preservation factors or techniques ('hurdles') in order to achieve multi-target, mild but reliable preservation effects. Attractive applications have been identified in many food areas. The microbial stability and safety of most traditional and novel foods is based on a combination of several factors (hurdles), which should not be overcome by the microorganisms present. This is illustrated by the so-called hurdle effect. The hurdle effect is of fundamental importance for the preservation of foods, since the hurdles in a stable product that controls the microbial spoilage, food-poisoning and, in some instances, the desired fermentation process. Hurdle technology is a method of ensuring that pathogens in food products which can be eliminated or controlled. This means the food products will be safe for consumption, and their shelf life will be extended. Hurdle technology usually works by combining more than one approach. These approaches can be thought of as "hurdles" the pathogen has to overcome if it is to remain active in the food. The right combination of hurdles can ensure all pathogens are eliminated or rendered harmless in the final product.

Hurdle technology has been defined by Leistner as an intelligent combination of hurdles which secures the microbial safety and stability as well as the organoleptic and nutritional quality and the economic viability of food products. The organoleptic quality of the food refers to its sensory properties that are its appearance, taste, smell and texture. The hurdle concept illustrates only the well-known fact that complex interactions of temperature, water activity, pH, redox potential, etc. are significant for the microbial stability of foods. Examples of hurdles in a food system are high temperature during processing, low temperature during storage, increasing the acidity, lowering the water activity or redox potential, or the presence of preservatives. According to the type of pathogens and how risky they are, the intensity of the hurdles can be adjusted individually to meet consumer preferences in an economical way, without compromising the safety of the product. Hurdle technology is used in industrialized as well as in developing countries for the gentle but effective preservation of foods.

Previously hurdle technology, i.e., a combination of preservation methods, was used empirically without much knowledge of the governing principles. Since about 20 years the intelligent application of hurdle technology became more prevalent, because the principles of major preservative factors for foods (e.g., temperature, pH, a_w, E_h, competitive flora), and their interactions, became better known. Recently, the influence of food preservation methods on the physiology and behaviour of microorganisms in foods, i.e. their homeostasis, metabolic exhaustion, stress reactions, are taken into account, and the novel concept of multi-target food preservation emerged. In the present contribution a brief introduction is given on the potential hurdles for foods, the hurdle effect, and the hurdle technology. However, emphasis is placed on the homeostasis, metabolic exhaustion, and stress reactions of microorganisms related to hurdle technology, and the prospects of the future goal of a multi-target preservation of foods.

There can be significant synergistic effects between hurdles. For example, gram-positive bacteria include some of the more important spoilage bacteria, such as *Clostridium, Bacillus* and *Listeria*. A synergistic enhancement occurs if nisin is used against these bacteria in combination with antioxidants, organic acids or other antimicrobials. Combining antimicrobial hurdles in an intelligent way means other hurdles can be reduced, yet the resulting food can have superior sensory qualities.

Possible Food Contaminants in Tea
Aflatoxins

Aflatoxins are a group of mycotoxins produced by *Aspergillus flavus* and *Aspergillus parasiticus*. These may grow in plant leaves, nuts, dried fruit and infect stored cereal grains where they produce the aflatoxins. When such toxins are formed they do not go away. They are heat stable; thus stay in the food along the food chain, unaffected by heat treatments such as pasteurization. When a cow eats a feed contaminated with aflatoxin B1, the activity of the cow changes the aflatoxin B1 to aflatoxin M1, which ends up in the milk.

The toxicity of aflatoxin is mainly due to its carcinogenicity. This is because aflatoxins are genotoxic, meaning it affects the genetic material. Genotoxins have a direct dose-response relationship, so they do not have a threshold dose to exceed before they have effect. Thus, there is no tolerable daily intake (TDI) for aflatoxins, which are to be kept at a level as low as possible. Though, maximum levels are set in the European Union, for example 0.05 µg/kg of milk for aflatoxin M1.

Mycotoxins

Mycotoxins are produced by mold, and there is a range of different compounds originating from different types of mold among the Deuteromycetes. The Mycotoxins are secondary metabolites, toxic in low concentrations in vertebrates.

Mycotoxins in the product can only be prevented at the source where the mold infection occurs, since the toxins are impossible to remove. In the case of made tea, control measures could be taken at the production and handling.

Except for the opinions, there are no proper examples available to prove aflatoxin contamination in made tea. Thus, there is a very minimal risk for getting the aflatoxins into the made tea. Since aflatoxins have a cumulative negative effect in humans they should be considered as hazards in the HACCP plan of the food safety management system.

Salmonella spp.

Salmonella spp. is a gram-negative, facultatively anaerobic organism, which does not form spores. Growth occurs at 5 – 47°C, and the organism is heat sensitive. It is a zoonotic organism, which may be found in different animals' guts.

Because *Salmonella* may be found in animals, it has a good chance of contaminating foodstuff of animal origin, like meat, milk and egg. For an example, Sweden and Finland there is zero-tolerance regarding *Salmonella* infection among domestic animals, which lowers the risk of human infection. Good handling and heat treatment of the food is necessary to decrease the risk.

Salmonella may cause either enteritis or a systemic infection. The gastrointestinal infection has symptoms as milk fever, vomiting and diarrhoea lasting for a few days up to more than a week. The infectious dose is in the order of 10^6 cells, but in some cases it has been much lower than that. The systemic infection is caused by invasive, host-adapted serotypes. The bacteria then spread in the body and causes fever, headache and diarrhoea. *Salmonella* infection may be fatal.

The tea leaves are plucked and heaped on the floor or in the field and transported to the factory with high risk of contamination form humans due to handling. The raw material further directly handled by many operators during withering, rolling, fermentation, sifting and packing, which increase the risk of *Salmonella* infection. Hence, there is a risk for *Salmonella* contamination in the product. Luckily the organism's heat sensitivity makes it possible to eliminate with LTHT treatment. The severity of the disease in combination with the risk makes *Salmonella* a hazard to target in the HACCP plan.

Escherichia coli

Escherichia coli are in the family Enterobacteriaceae, gram negative, rod shaped, non-spore forming, and motile or non-motile. They can grow under aerobic and anaerobic conditions where they grow best at 37°C. Therefore, it is easy to eradicate by simple boiling or basic sterilization.

E. coli O157:H7 is a well-studied strain of the bacterium *E. coli*, which produces Shiga-like toxins, causing severe illness. *E. coli* is transmitted to humans primarily through consumption of contaminated foods, faecal contamination of water and other foods. Infectious dose of *E. Coli* is 10^6 - 10^8 logs of organisms.

The tea leaves are plucked and heaped on the floor or in the field and transported to the factory with high risk of contamination or cross contamination from humans due to handling. The raw material further directly handled by many operators during withering, rolling, fermentation, sifting and packing, which increase the risk of *E. coli* infection. Hence, there is a risk for *E. coli* contamination in the product due to poor personal hygiene and improper cleaning of utensils. Luckily the organism's heat sensitivity makes it possible to eliminate with LTHT treatment. The severity of the disease in combination with the risk makes *E. coli* a hazard to target in the HACCP plan.

Staphylococcus aureus

Staphylococcus aureus is a facultative anaerobic organism, which does not form spores. Growth occurs at 7 – 45°C with optimum growth at 37°C, and the organism is heat sensitive. The presence of *Staphylococcus aureus* is of concern in products that are fermented. *Staphylococcus aureus* can multiply to high numbers during fermentation if the product is not rapidly fermented (e.g., the starter culture is not active) and cause a toxin to be produced that can cause illness to consumers.

The bacteria are common in the environment and are often found on skin, nose, mouth or boils and cuts of people. The product may generally become contaminated with *Staphylococcus aureus* from the raw materials or from human contact.

Generally, it takes high numbers and growth of *Staphylococcus aureus* to cause a hazard with a medium dose. The symptoms of *Staphylococcus aureus* food poisoning are nausea, vomiting, stomach cramps, prostration, diarrhoea and last for 6 to 24 hours.

Proper cooking, fermentation, cooling, storage and personal hygiene of food handlers can prevent growth while minimizing cross contamination of *Staphylococcus aureus* and more importantly, the production of their toxins. However, cooking will not destroy toxins once they are formed in food.

The tea leaves are plucked and heaped on the bare floor or in the field and transported to the factory with high risk of contamination form humans due to handling. The raw material further directly handled by many operators during withering, rolling, fermentation, sifting and packing, which increase the risk of *Staphylococcus aureus* contamination..

Further, the intermediate product is fermented for minimum 2.5 hours and may be as long as 6 hours in some specific tea grades. The severity of the disease in combination with the risk makes *Staphylococcus aureus* a hazard to target in the HACCP plan.

Pesticide Residues
A pesticide is any product that kills or controls various types of pests where a pest is defined as a plant or animal that is harmful to man or the environment. Most of us recognize that certain insects, weeds, and rodents that are pests, but the use of pesticides is not limited to the control of these pests, where other harmful pests can include birds, snails, fungi, algae, and bacteria which needs specific concerns while controlling them. Inability to control pests has had a tremendous impact on world history, i.e. Millions of people died from bubonic plague (the infamous Black Death) before it was discovered that rat fleas carried the disease. A similar number have died from malaria, which is transmitted by mosquitoes. The Irish potato famine in the 1850s was caused by a fungus that destroyed that country's major food source, resulting in many deaths and prompting mass migrations to the United States. Today, bubonic plague is of little concern, the potato fungus is an insignificant problem, and malaria has been greatly reduced in the world. The use of pesticides is the primary reason these problems are no longer a threat.

But pesticides are also potentially toxic to humans, which may induce adverse health effects including cancer, immune or nervous systems and it may cause effects on reproduction. Thus, pesticides need proper investigations before it can be authorized for use where pesticides should be tested for all possible health effects and the results should be analyzed by experts to assess any risks to humans. Scientific studies of the potential health effects of hazardous chemicals, such as pesticides, allow them to be classified as carcinogenic (can cause cancer), neurotoxic (can cause damage to the brain), or teratogenic (can cause damage to a fetus). This process of classification, called "hazard identification," is the first step of "risk assessment". These problems causing compounds after usage are called pesticide residues.

Thus pesticide residues are the deposits of pesticide's active ingredient, its metabolites or breakdown products present in some component of the environment after its application, spillage or dumping. Residue analysis provides a measure of the nature and level of any chemical contamination within the environment and of its persistence. It is often difficult to correlate pesticide residues in the environment with effects on fauna and/or ecological processes. They can, however, show whether an animal or site has been exposed to chemicals and identify the potential for future problems. All pesticides are subject to degradation and/or metabolism once released into the environment. The rates of degradation and dissipation vary greatly from pesticide to pesticide and situation to situation.

Type of Pesticides
There are various kinds of pesticides in use based on the type of pest and intensity of the pest attack which can be summarized as:

Organochlorines
Mobility of organochlorines in soil is generally limited; although it is greater in sandy soil. They tend to be bound in clay soils with limited leaching. Residues of the parent compound or metabolites can be found in soil, sediment, vegetable samples and in vertebrates/invertebrates for extended periods. Their solubility in water is low, although residues can be detected in water where there is extreme contamination and, particularly, on suspended matter in water.

E.g. lindane (gamma isomer of benzene hexachloride), dieldrin, DDT (p-p′ isomer), heptachlor, endosulfan

Organophosphates
Organophosphates have a fairly limited environmental persistence and residues in living specimens generally are not detected, or only as metabolites in specific cases. Water solubility is

variable but higher than with the organochlorines; residues generally break down quite quickly in water (hydrolysis) which is not generally detected except where the contamination is quite recent. Soil residues are similarly short-lived. Residues are probably only of interest for 5–15 days after spraying unless in shaded areas or where the concentrations applied are high.
E.g. fenitrothion, fenthion.

Carbamates

Residues of parent compounds are generally not environmentally persistent; metabolites are rapidly excreted by vertebrates where w, Water solubility is moderate; greater for the metabolites. Most carbamates are relatively stable in water of neutral pH. Stability and mobility in soil varies between compounds. Environmental residues are probably only of interest for 10–20 days after spraying, although in certain soils and in water, extended monitoring may be required.
E.g. aldicarb, carbaryl, propoxur.

Pyrethroids

Pyrethroids are insecticides and generally non-persistent in the environment. It is being rapidly degraded in the presence of strong sunlight, where residues are probably only of interest for 5–7 days after spraying, unless in shaded areas and where the concentrations applied are particularly high. Proper and accurate detection of residues requires a specialist laboratory.

E.g. Cypermethrin, permethrin, deltamethrin.

Insect Growth Regulators

Benzoyl urea IGRs generally act by inhibition of chitin synthesis and moulting, thus interfering with the formation of the insect cuticle. They are increasingly used for the control of leaf-eating insects (mandibulate herbivores) in forestry, ornamentals and fruit. Their low water solubility and adsorption by soil reduces their environmental impact and in general use, residues are only likely to be detected in soil. There may be some, limited non-target effects in treated areas. There are also IGRs which act as juvenile hormone mimics, disrupting or preventing maturation of immature invertebrates.

E.g. diflubenzuron – benzoyl urea IGR, teflubenzuron – benzoyl urea IGR, triflumuron – benzoyl urea IGR, methoprene – terpenoid IGR (juvenile hormone mimic) and fenoxycarb – bridged diphenyl carbamate IGR (juvenile hormone analogue).

Herbicides

Although of relatively low acute toxicity to most animals, herbicides can indirectly affect a variety of species through the removal of vegetative cover. Environmental persistence of the herbicides varies; some are readily absorbed by and degraded in soil (e.g. paraquat) whilst others are more persistent and, with relatively high water solubility, considered to be quite mobile (e.g. triazine materials) where residues transferring (leaching) to waterways are a recognized problem. Residues in wildlife are generally transient with rapid metabolism and excretion. The significance of residues depends upon the applied material e.g. with 2,4-D, residues decline quite quickly with a half-life of <7 days in soil; with the triazine herbicides or with products such as linuron/diuron, persistence is considerably greater and residues can be present for months. The persistence of sulphonyl urea herbicides varies although at the extremely low rates they are applied under normal use, the residues present are particularly low and the analysis can be difficult.

E.g., 2-4-D, atrazine, linuronm, chlorsulfuron

Fungicides

Some fungicides can have adverse environmental effects but, although they are used extensively in the field for cereal production, their use patterns suggest limited scope for environmental contamination except as the result of disposal (e.g. from large-scale dip treatment operations) or accidental contamination (spillage, etc.). Water solubility and stability are variable; some fungicide residues can be detected in water for periods of days through to months.

E.g. carbendazim, chlorothalonil, metalaxyl

Soil Fumigants

Materials such as methyl bromide (use now heavily restricted under the Montreal Protocol) and 1,3-dichloropropene are examples of materials used as soil fumigants. Under controlled use, soil fumigants do not pose a substantial environmental problem unless they are allowed to contaminate watercourses (methyl bromide is highly soluble in water, 13.4 g l⁻¹ at 25 °C, 1,3-dichloropropene is less soluble, 2 g l⁻¹ at 20 °C). The materials are volatile and dissipate to atmosphere on aeration of the soil.

The use of toxic pesticides to manage pest problems has become a common practice around the world. Pesticides are used almost everywhere, not only in agricultural fields, but also in homes, parks, schools, buildings, forests, and roads. It is difficult to find somewhere where pesticides aren't used from the can of bug spray under the kitchen sink to the airplane crop dusting acres of farmland; our world is filled with pesticides. In addition, pesticides can be found in the air we breathe, the food we eat, and the water we drink.

The presence of minute residues of pesticides in food has caused some people to ask, "Is our food supply safe? "due to the recently increased attention on chemical residues in food. Current evidence strongly indicates that US foods are safe where Food and Drug Administration (FDA) officials recently stated that "pesticide residues occurring in foods in the U.S. pose a very minor if not negligible risk to public health." However, public perceptions of risks from pesticides differ markedly from this official viewpoint and the product that we buy everyday are not coming from USA and they also differ from actual risks attributable to these products.

Pesticides and Human Health
Pesticides have been linked to a wide range of human health hazards, ranging from short-term impacts such as headaches and nausea to chronic impacts like cancer, reproductive harm, and endocrine disruption. Acute dangers - such as nerve, skin, and eye irritation and damage, headaches, dizziness, nausea, fatigue and systemic poisoning - can sometimes be dramatic, and even occasionally fatal. Chronic health effects may occur years after even minimal exposure to pesticides in the environment, or result from the pesticide residues which we ingest through our food and water. A July 2007 study conducted by researchers at the Public Health Institute, the California Department of Health Services, and the UC Berkeley School of Public Health found a six fold increase in risk factor for autism spectrum disorders (ASD) for children of women who were exposed to organochlorine pesticides.

Pesticides can cause many types of cancer in humans where some of the most prevalent forms include leukemia, non-Hodgkins lymphoma, brain, bone, breast, ovarian, prostate, testicular and liver cancers. In February 2009, the Agency for Toxic Substances and Disease Registry published a study that found that children who live in homes where their parents use pesticides are twice as likely to develop brain cancer versus those that live in residences in which no pesticides are used. Studies by the National Cancer Institute found that American farmers, who in most respects are healthier than the population at large, had startling incidences of leukemia, Hodgkins disease, non-Hodgkins lymphoma, and many other forms of cancer.

There is also mounting evidence that exposure to pesticides disrupts the endocrine system, wreaking havoc with the complex regulation of hormones, the reproductive system, and embryonic development. Endocrine disruption can produce infertility and a variety of birth defects and developmental defects in offspring, including hormonal imbalance and incomplete sexual development, impaired brain development, behavioral disorders, and many others. Examples of known endocrine disrupting chemicals which are present in large quantities in our environment include DDT (which still persists in abundance more than 20 years after being banned in the U.S.), lindane, atrazine, carbaryl, parathion, and many others.

Multiple Chemical Sensitivity (MCS) is a medical condition characterized by the body's inability to tolerate relatively low exposure to chemicals. This condition, also referred to as Environmental Illness, is triggered by exposure to certain chemicals and/or environmental pollutants. Exposure to pesticides is a common way for individuals to develop MCS, and once the condition is present, pesticides are often a potent trigger for symptoms of the condition. The variety of these symptoms can be dizzying, including everything from cardiovascular problems to depression to muscle and joint pains. Over time, individuals suffering from MCS will begin to react adversely to substances that formerly did not affect them. For individuals suffering from MCS, the only way to relieve their symptoms is to avoid those substances that trigger adverse reactions. For some individuals, this can mean almost complete isolation from the outside world.

Heavy Metals

Although there is no clear definition of what a heavy metal is, density is in most cases taken to be the defining factor. Heavy metals are commonly defined as those having a specific density of more than 5 g/cm^3. Heavy metals have been used in many different areas for thousands of years. Lead has been used for at least 5000 years, early applications including building materials, pigments for glazing ceramics, and pipes for transporting water. In ancient Rome, lead acetate was used to sweeten old wine, and some Romans might have consumed as much as a gram of lead a day. Mercury was allegedly used by the Romans as a salve to alleviate teething pain in infants, and was later (from the 1300s to the late 1800s) employed as a remedy for syphilis. Claude Monet used cadmium pigments extensively in mid 1800s, but the scarcity of the metal limited the use in artists' materials until the early 1900s.

Although several adverse health effects of heavy metals have been known for a long time, exposure to heavy metals continues, and is even increasing in some parts of the world, in particular in less developed countries, though emissions have declined in most developed countries over the last 100 years. The main threats to human health from heavy metals are associated with exposure to lead, cadmium, mercury and arsenic (arsenic is a metalloid, but is usually classified as a heavy metal). Emissions of heavy metals to the environment occur via a wide range of processes and pathways, including the air (e.g. during combustion, extraction and processing), to surface waters (via runoff and releases from storage and transport) and to the soil and hence into ground waters as well as crops. Atmospheric emissions tend to be of greatest concern in terms of human health, both because of the quantities involved and the widespread dispersion and potential for exposure that often ensues.

9

A Case Study on Food Safety

Application of Food Safety Management Systems in Sri Lankan Orthodox Black Tea Industry

A study was carried out to understand the food safety issues in the tea manufacturing process with regard to the ISO 22000 requirements while identifying the major gaps pertaining to the food safety. Hence, this research study was focused to introduce well developed ISO 22000 generic model for tea manufacturing process, with a view to cover the gaps, identified by the Gap Analysis. The system, was intend to be introduced was gauged by conducting Internal Audit by resulting user friendly paper based model along with required documents and formats which can be customize according to customer. This will include formats, examples, presentations, check lists, reports and notes to support tea industrialists and consultants. In addition to that, it was mandatory to visit and interview some auditors, industries and consultants in the respective field in order to get more accurate and consistent data for the betterment of the generic model.

The prepared data was analyzed and published for the betterment of industries and consultants as guiding materials when and where they required. The research was also focused to develop equipment, qualify in order to bridge the gaps already identified in certain areas of tea manufacturing process. The following case study represents the results of above study.

The effectiveness of the application of HACCP based FSMS in assuring food safety, has been proven in many industries around the world, where HACCP improves business performance [123], [126], [127] operational performance [123] and overall quality performance [67]. According to recent research findings on food safety violations in tea industry, following table (9.1) describes local industry's position based on the GMP requirements mentioned in food act as well as tea board guidelines.

Table 9.1: Descriptive statistics - O&MR, ED&F, SF, DF, Cleaning, PCS, PH, QAS, Total

Variable	Mean	StDev	Minimum	Median	Maximum
O&MR	6.767	1.394	4.000	7.000	9.000
ED&F	17.767	4.247	11.000	18.000	25.000
SF	5.140	1.754	3.000	5.000	9.000
DF	5.930	1.298	4.000	6.000	8.000
Cleaning	3.047	0.722	2.000	3.000	4.000
PCS	2.442	0.666	2.000	2.000	4.000
PH	3.326	0.919	2.000	3.000	6.000
QAS	11.884	2.666	6.000	12.000	16.000
Total	56.72	11.87	39.00	56.00	78.00

The summary statistics showed as above (table 9.1) Mean, Median, Standard Deviation, Minimum and Maximum values for each variable or the areas segregated according to the importance of GMP

~ 198 ~

requirements and the weightages given considering the impact on food safety. Based on the above results, tea factories have implemented GMP with a total median of 56 where minimum and maximum values varied in between 39~78 within the sample. In addition, Spearman Rho was (as to the below table 9.2) provided relevant P-values which further explains the correlations between different components of testing criteria.

Table 9.2: Spearman rho - O&MR, ED&F, SF, DF, Cleaning, PCS, PH, QAS, Total

	O&MR	ED&F	SF	DF	Cleaning	PCS	PH	QAS
ED&F	0.794 0.000							
SF	0.692 0.000	0.775 0.000						
DF	0.522 0.000	0.592 0.000	0.769 0.000					
Cleaning	0.584 0.000	0.570 0.000	0.582 0.000	0.419 0.005				
PCS	0.482 0.001	0.415 0.006	0.268 0.082	0.071 0.653	0.539 0.000			
PH	0.615 0.000	0.627 0.000	0.503 0.001	0.465 0.002	0.405 0.007	0.124 0.429		
QAS	0.735 0.000	0.788 0.000	0.762 0.000	0.639 0.000	0.731 0.000	0.494 0.001	0.508 0.001	
Total	0.853 0.000	0.929 0.000	0.874 0.000	0.736 0.000	0.704 0.000	0.469 0.002	0.654 0.000	0.912 0.000

Cell Contents: Spearman rho
P-Value

As to the n coefficient or Spearman's rho, relationships between the 8 different areas (O&MR, ED&F, SF, DF, Cleaning, PCS, PH, QAS,) of the GMP; were showed as above table 9.2. As to two-tailed test of significance; there was significant positive relationship between the organization and management responsibility (O&MR), establishment design and facilities (ED&F) which was the strongest correlation among tested components. Nevertheless, it further shows that the most insignificant or weakest relationships can be observed among distribution facilities (DF) and personal hygiene (PH) with pest control systems (PCS). On the other hand, quality assurance systems had stronger or moderate relationship with all the other components tested, where better the quality assurance system; better their good manufacturing practices (GMP) and same pattern can be observed in O&MR. All un-circled results were showed moderate relationships between tested components.

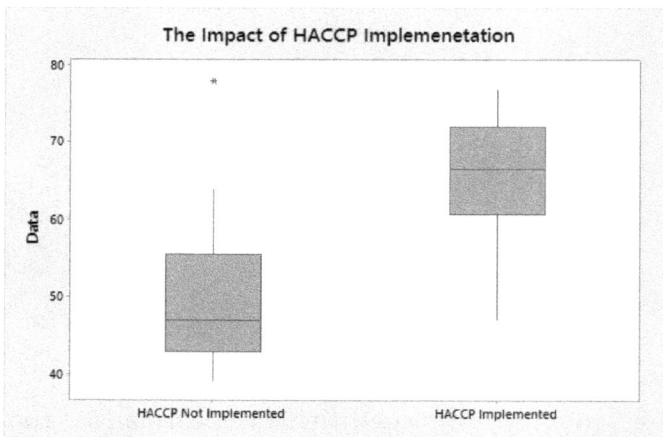

Figure 9.1: The impact of HACCP implementation on food safety

The above box plot compare of tea factories with and without HACCP system based on the same research above explained with combination of nonparametric Mann-Whitney test. The above figure (9.1) shows the impact of HACCP based FSMS with higher levels of GMP and food safety achievement between the two groups due to the implementation of HACCP. Considering the group of factories used for evaluation, the G_0 (Factories without HACCP) group has lower median value (47) with short whisker (39) for the lower side and longer whisker for upper side (64) with an outlier (78). In contrast, factories implemented HACCP based FSMS (G_1 - Factories with HACCP) had a higher median value (66.5) with the upper whisker, which is short (77) and lower whisker is long (47). The data set spread with left skewed pattern in HACCP implemented factories while G_0 had a right skewed pattern of data showing that there was significant difference between two groups based on the availability of HACCP based FSMS or not, which further confirms that; HACCP based FSMSs such as HACCP improve the business performance [123], [126], [127] which is also true with ISO 22000 [128].

As to the nonparametric Mann-Whitney test, resulting medians are given in the Table 9.3, the table further shows the relevant W and P-values related to each component tested under GMP requirements (OM&R, ED&F, SF, DF, Cleaning, PCS, PH, QAS) as well as the overall results for impact of HACCP implementation (Total). Considering the overall achievements due to the implementation of HACCP based FSMS, the medians were 47.00 and 66.50, where there was a

95.2% confidence interval for the difference in population medians (HACCP Not Implemented - HACCP Implemented) is [-22.003 to -10.997]. The test statistic W = 376.5 has a p-value of 0.0000 and 0.0000 when adjusted for ties. Since the p-value is less than the chosen α level of 0.05, there is sufficient evidence to reject H_0. Thus, the data supports the hypothesis that there is a difference between the population medians and it further conclude that; implementation of HACCP based FSMS has a positive impact of improving food safety assurance in Sri Lankan tea industry whereas Implementation of QAS are improving operational efficiency and reduce the production costs [129], [130], [131] and being recognized as employing "best practice", or a "good system", with new technology and innovation [129], [132], [122], [131].

In contrast, table 9.3 further shows that, there is no any significant relationship between HACCP based FSMS implemented factories and not implemented factories on cleaning (P-value 0.074) and pest control systems (P-value 0.619) where chosen α level was higher than the 0.05. Considering the current applications in Sri Lankan tea industry, it is very true, because both category of factories had similar levels of cleaning with insufficient documentations and very low level of pest control system implementations where table 03 shows (Red colour circles) insignificant difference at 95% confidence level. Other P-values shows (blue colour circles) that

HACCP based FSMS implementation has helped the tea industry to efficiently implement GMP compliances while improving the organizational management and responsibility, establishment design and facilities, storage facilities, distribution facilities, personnel hygiene and quality assurance systems at 95% confidence interval.

Table 9.3: Impact of implementation of HACCP based FSMS on food safety assurance in Sri Lankan tea industry

Variable	HACCP	Mean	StDev	Minimum	Median	Maximum	W	P-Value
O&MR	0	6.160	1.281	4.000	6.000	9.000	414.0	0.0004
	1	7.611	1.092	5.000	8.000	9.000		
ED&F	0	15.280	3.273	11.000	15.000	25.000	369.0	0.0000
	1	21.222	2.777	15.000	22.000	24.000		
SF	0	4.440	1.557	3.000	4.000	9.000	424.0	0.0010
	1	6.111	1.568	3.000	6.000	8.000		
DF	0	5.480	1.262	4.000	5.000	8.000	441.0	0.0380
	1	6.556	1.097	5.000	7.000	8.000		
Cleaning	0	2.800	0.645	2.000	3.000	4.000	450.5	0.0740
	1	3.389	0.698	2.000	3.500	4.000		
PCS	0	2.280	0.542	2.000	2.000	4.000	487.0	0.6190
	1	2.667	0.767	2.000	2.500	4.000		
PH	0	3.120	1.013	2.000	3.000	6.000	465.0	0.0188
	1	3.611	0.698	3.000	3.500	5.000		
QAS	0	10.360	2.177	6.000	10.000	15.000	371.5	0.0000
	1	14.000	1.645	10.000	14.000	16.000		
Total	0	49.92	9.35	39.00	47.00	78.00	376.5	0.0000
	1	65.17	7.90	47.00	66.50	77.00		

Note: 0 - HACCP Based FSMS Not Implemented,
1 - HACCP Based FSMS Implemented

Organization and Management Responsibility

According to the Table 1, general population of factories was half way through O&MR achievements. The achievements can be varied within minimum and maximum values 04 - 09 with a median in between 6.0 - 7.0 at 95% confidence level. In addition, the P-value and W (0.0004, 414) for nonparametric Mann-Whitney test further justify the alternative hypothesis, which was implementation of HACCP based FSMS has positively impacted (table 9.3) on tea industry in improving GMP implementation and there by food safety of the final product; due to top management's support and influence on adoption of QASs [133], [134].

Establishment Design and Facilities
The Establishment Design and Facilities had a great impact on the entire food safety of the facility as well as all the components tested on GMP. The summary statistics shows 95% confidence interval (table 9.1) for minimum of 11 and maximum of 25 and mean value of 17.79 for the sample which can be used as an indication of the achievements in GMP implementation. This further explains the improvements required. The P-vale, 0.0000 and W, 369 where difference between mean values were 15.280 and 21.222 (table 9.3) for the nonparametric Mann-Whitney test which further explains the positive impact of implementation of HACCP based FSMS in tea industry.

Storage Facilities
According to the table 9.1, the study sample showed mean of 4.59 to 5.67 and median of 4 - 6 under 95% confidence interval (CI) which further explain the improvements needed due to the high number of 3

and 4 scores achieved out of 10, as to the results obtained. When compared the factories with and without HACCP based FSMS, mean was 4.44 for factories without HACCP and 6.11 for factories with HACCP based FSMS, where W - 424 and P-value - 0.001 (table 9.3) which shows the relationship between implementation of HACCP based FSMS and its positive impact generated in tea industry.

Distribution Facilities

As to the results, most of the tea factories involved in the evaluation has achieved 5 or more than 5 marks and according to the curve it is right shifted with a median of 6 and mean of 5.93 where the industry has better distribution and collection facilities generally under a 95% confidence level. Nevertheless, DF has moderately positive relationship (table 9.3) where W 441 and P-value was 0.0380 which still rejects the null hypothesis while proving that, HACCP based FSMS implementation had improved distribution facilities of the tea industry at a 95% CI. DF has a strong correlation with storage facilities while it has no relationship with PC according to the Table 3, whereas it further showed a moderately high relationship with QAS concluding that distribution facilities are moderately improved than other areas of tea industry.

Cleaning

Out of the 5 marks given most of the factories have achieved 3 marks as to the results. The median was 3 where mean value also changed within a very narrow range. The minimum and maximum was changed in-between 2 - 4; which shows that, tea industry needs to improve their cleaning operation as a whole. According to the Spearman's rho, the cleaning has a stronger relationship with QAS (table 9.3), but it further showed that there was no any significant improvements in cleaning whether factory has implemented a HACCP based FSMS or not. This was the visual observation too, since tea industry need to focus on its cleaning activities because cleaning was very inadequate in many places without making any impact due to the implementation of HACCP based FSMS.

Pest Control Systems

As to the results, most of the tea factories achieved 2 points out of 5 which shows very low levels of achievements in tea industry. Accordingly, mean varies between 2.0 - 2.3 under 95% confidence interval, where minimum and maximum varies between 2 - 4. In addition, PCS has no relationship with SF (P, 0.082) or DF (P, 0.653) according to Table 2, while it has very weak relationship with PH. Furthermore, PCS showed null hypothesis was true where HACCP based FSMS has not made any differences to the improvement of GMP through improving PCS. Thus tea industry has to focus on improving PCS for the betterment of food safety.

Personal Hygiene

The statistical summary (table 9.1) on personnel hygiene also shows the very lower achievements by concentrating the mean value of 3.3 which was far below the average value. Since minimum and maximum values were 2 to 6, the achievements in the general factory population also show the same pattern at 95% confidence. In line with the Spearman rho (table 9.2), PH has moderate relationships with all the components evaluated other than PCS which was discussed previously. However, PH has accepted alternative hypothesis (W, 465 and P, 0.0188) where it was positively correlated, but as to the mean and the median PH achievements were very low for both HACCP based FSMS implemented tea factories as well as not implemented. Thus tea industry has to consider improving its operators PH for the betterment of consumer safety.

Quality Assurance Systems

According to the table 9.1, it shows the results of quality assurance systems (QAS) implementation of the sample evaluated. Based on the above statistical outputs, it is obvious that industry needs a shift in quality assurance system implementation. Quality assurance systems had improved the operational performance [123] with food safety and quality of product manufactured [67] while improving the infrastructure which intern had increased the points scored by individual factories who has implemented HACCP based FSMS. Out of given 20 points, the result has varied between minimum of 6 to maximum of 16 where mean value has shifted only from 11.0 to 12.7 proving that QAS has more works to be done in the tea industry. As to the Spearman rho, QAS has strong relationship with O&MR, ED&F, SF and Cleaning while it has moderate relationships with all other areas under consideration. All the P-values were much less than 0.05 proving that the alternative hypothesis was acceptable. Besides, mean value and median (table 9.3) had considerable difference between factories with HACCP based FSMS and without. It also showed a positive correlation while accepting the alternative hypothesis where HACCP based FSMS implementation in tea industry has created an enabling environment to improve the GMP conditions as well as food safety in tea industry.

In addition, 5S implementation has further help to systematize and improve the effectiveness of implementation of GMP where most of the factories which has implemented HACCP based FSMS has implemented 5S before they implemented FSMS. The efficiency of processing, recording and personnel hygiene were satisfactorily improved in factories who have implemented HACCP based FSMS and 5S, where improving product quality is an important motivation to adopt HACCP by firms in the agri-food sector [132] Japanese 5S has played a major role in improving infrastructure and training of workers where visual implementations were highly concerned. Nevertheless, factories with HACCP based FSMS has better infrastructure and systematic operations with trained operators rather than factories without any HACCP based FSMS, because food producers adopt HACCP in order to satisfy downstream customers in the food chain [135], [136].

Food safety violations in low grown orthodox black tea manufacturing industry were further evaluated using Pareto analysis which was based on 20:80 principle and pie chart. After the gap analysis, the collected information was used to find out major issues related to food safety violations and the generic documentation system development was initiated to fill the gaps in documentation. The relevant requirements were carefully defined according to the documentation requirements stated in ISO 22000:2005. In addition, ISO 30300:2011 was considered for the development of record management system while merging 5S into the ISO 22000:2005 within the area of work instructions, cleaning and preventive maintenance [137]. The tea industry was carefully analyzed with reference to the operating environment and the type of documents needed for the relevant processes compatible with ISO 22000:2005 FSMS to ensure the proper management of the food safety.

After conducting the gap analysis for the industry sector; the major areas for the interventions needed were decided based on Pareto principle where 20% of the root causes were responsible for the 80% of the problems which was shown in figure 9.2 below. Thus study was planned to address most critical issues of the industry with regards to the food safety violations and the food safety assurance. According to the Pareto chart (figure 9.2), it was obvious that Establishment Design and Facilities was the major root cause for the food hygiene problems identified where continuous attention and Top Management commitment as well as additional capital investments were needed to improve design and facilities of the manufacturing plants in the sector.

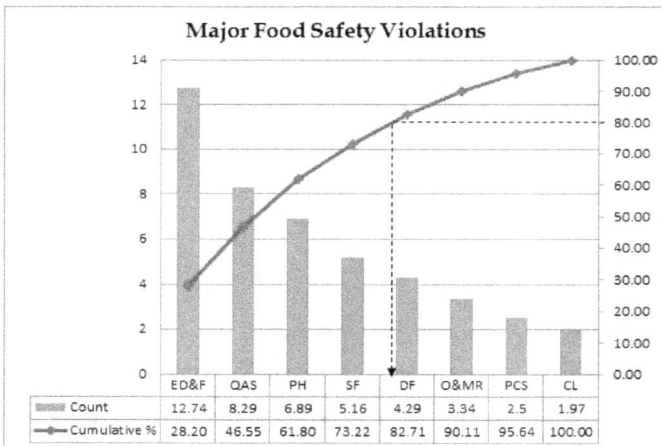

Major Food Safety Violations

	ED&F	QAS	PH	SF	DF	O&MR	PCS	CL
Count	12.74	8.29	6.89	5.16	4.29	3.34	2.5	1.97
Cumulative %	28.20	46.55	61.80	73.22	82.71	90.11	95.64	100.00

Figure 9.2: Pareto analysis on food safety violations

On the other hand, next major contribution for food safety violation was found in the area of Quality Assurance Systems (figure 9.2), where none of the existing systems were in compliance with complete food safety. However, firms strive to maintain or improve both safety and quality attributes together and such efforts are closely interrelated and most likely are managed as the whole in practice [138]. This was mostly due to the incomplete system developments, lack of expert knowledge in the industry and also due to the inappropriate practices.

Nonetheless, Personnel Hygiene also contributed to food safety violations because employee training was very rare and it was neglected even among the organizations having ISO 22000 certifications, where worker knowledge on this area was highly unsatisfactory. Evidences showed that conducted training programs were also not up to the standards and basically concentrated on very few areas of the requirements. Storage Facilities became the next contender, which was a part of Establishment Design and Facilities because if these facilities were adequately developed storage facilities were also developed according to the requirements. Furthermore, the fermentation area was identified as another major point in food safety violations where food hygiene was very vulnerable and ignored by many manufacturers including factories with HACCP based FSMSs. The gap was bridged by designing a fermentation trolley with specific features.

Even through Organization and Management Responsibility, Distribution Facilities and Cleaning was out of the 80% of Pareto Chart (figure 9.2) management responsibility was one of the major requirements because, if the management committed to improve the factory condition all these problems were eliminated. Thus there was a direct relationship between all other factors with Organization and Management Responsibility and it was a proportional relationship which is also true for the quality assurance systems. Cleaning can be improved with proper training and organizing workers in to area specific teams under the proper guidance of the management.

GMP Compliance in Black Tea Manufacturing

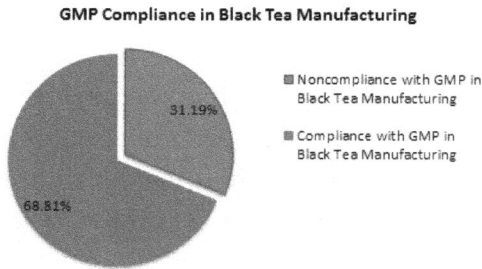

Figure 9.3: Comparison of GMP compliances

According to the above evaluations, ISO 22000 requirements were further analyzed to develop a generic model while absorbing existing practices into the generic model which can help tea manufactures with more convenient manner, since the manufacturing process is more or less similar with all orthodox black tea manufacturing facilities. The initiative was based on the allowances made by the ISO 22000:2005 which specifically states "this international standard allows an organization (such as a small and/or less developed organization) to implement an externally developed combination of control measures" [139] which was further stressed by the official guide book by ISO and the ITC "ISO 22000 - Are you ready?" where ISO 22000:2005 permits an organization to choose between developing its own control measures or using one developed by an industry association, government agency, university etc. [103].

Above pie chart (figure 9.3) shows the average compliance levels achieved in the area of GMP based on the reference sample, which further explain the 31.19% of GMP deficiencies that must be improved before planning an ISO 22000:2005 food safety management system. The generic model was a customizable food safety system based on ISO 22000:2005, which can be readjusted according to the factory specific requirements where very few things to be changed before implementation. The advantage was that, quality officers working in tea manufacturing facilities will be able to modify the system according to factory requirements for the initiation of the system and they also can conduct internal training programs based on the given presentations. If they have the opportunity to participate for externally conducted internal auditing programs, then they can ready the system for initial auditing, at the first stage audit or mock audit by the certification service provider will request for any additional specific requirement based on the specific manufacturing process. Auditor may even tell them what exactly is the compliance criteria to be met before certification, this will help the industry to implement their own systems where ownership of the system belong to the factory officer and he will motivate to implement and run the system while learning it, rather than getting consultancies from someone don't know exactly about the industry. Thus generic model was designed with above objectives where it contains all the mandatory documents which were described in ISO 22000:2005 standard.

5S and Ceylon Tea

The major problem encountered in the improvement was production and development of infrastructure, because cost of production (COP) was being highest among tea producing countries, whereas the profitability is comparatively less. The corporate sector has the highest COP which basically depends on the cost of green leaves and which in turn is dependent on the productivity of the field, wages, plucker intake, cost of other inputs such as weeding, fertilizing and transportation costs. The current rate of a green leaf 1kg was found to be around LKR 75.00 – 80.00 and plucking cost was about LKR 25.00 per 1kg. Accordingly, the total gross cost of the green leaf per 1kg was about LKR 60.00 where corporate sector has lower profitability rates due to the many other overheads accumulated while manufacturing process up to the auction. On the other hand, smallholder sector operates in different models where household labour is mostly utilized in many of the crop management and plucking operations as well as transport.

Nonetheless, consumer awareness on product safety is stronger. World food crises that prevailed in the past decades resulted doubts in the consumer's mind thereby causing lack of trust and confidence in products placed on the market. Fortunately, most of the companies have already given special attention on the product quality and consumer safety. A lot of good practices have been developed and implemented on a voluntary basis by manufacturers. These practices ensure achieving product safety satisfactorily [92]. Companies continuously challenge their internal quality systems and work on continuous improvement, thanks to new technologies and ways of working.

Considering these food safety problems and trade issues generated over the time, the International Standard Organization developed the ISO 22000 Food Safety Management System (FSMS) to harmonize the requirements of various food safety standards into integrated system while eliminating lots of trade issues faced on exports. The new standard ensures the complete food safety of entire food supply chain while satisfying global food safety statutory and regulatory requirements.

ISO 22000 is a quality assurance system introduced by ISO, to ensure consumer safety through food safety while eliminating trade issues, which is a further development of hazard analysis critical control point (HACCP) and other available food safety/quality assurance systems that ensures the food safety of entire food supply chain from farm to fork. ISO 22000 is a federative standard which harmonized the most of the food safety requirements set by different global standards and compatible with any food safety regulation worldwide.

ISO 22000 has been developed basically merging good manufacturing practices (GMP), HACCP and ISO 9001. Here the foundation layer is consist of GMP/GHP/GAP, Codex General Principles of Food Hygiene and Prerequisite programs which altogether creates very sound infrastructure and physical requirements to implement food safety requirements inside the plant focusing on basic food hygiene standards.

The total food safety is achieved through HACCP system of Codex Alimentarius using its seven principles to identify hazards and to control them under strict management plan. This includes the hazard analysis, identification of critical control points, establishment of critical control limits, monitoring procedures, corrective actions, record keeping and verification activities. However, these requirements are applied through mandatory food safety procedures. In addition, same procedures and activities are applied to the prerequisite programs and operational prerequisite programs identified according to the risk levels of the product manufactured.

The most effective food safety systems are established, operated and updated within the framework of a structured management system and incorporated into overall management activities of the organization concerned which provide the maximum benefits for interested parties. The standard integrates the HACCP system and application steps developed by Codex Alimentarius Commission. By means of auditable requirements, it combines the HACCP plan with (PRPs) perquisite programs [139]. The ISO 22000 FSMS has been developed based on risk based management model focusing the entire food supply chain through harmonization.

While improving the tea industry, Sri Lanka Tea board, Tea Research Institute and Plantation owners had committed greatly where various new technologies, improvements, polices and regulations introduced over the time and adapted. Some of these introductions were diminished over the time while others survived where few of them were greatly adapted/adsorbed by governing and regulatory organizations as well as plantation or factory owners and their subordinates. One of such introductions was Japanese 5S method which has helped tea industry to improve its productivity and organization of work place into a more productive system. Nevertheless, tea industry also adapted ISO 22000, HACCP, ISO 9001, UTC, Ethical Tea Partnership and many other different standards relating to buyer requirements or the company's individual preferences.

In current context, most organizations are look forward to find ways of reducing the cost, improve the quality while increasing the productivity in order to be more competitive and maintain the organization's excellent performance [140]. While considering ISO 22000 FSMS in contrast to Japanese 5S, it also has same areas of interventions which had more alike features where it was better to consider utilizing adapted systems instead of developing a completely new system. 5S was initially originated in Japan which dates back to the post developments of World War II [141], where it was used to improve the overall productivity of manufacturing through focusing cleanliness, orderliness and discipline with continuous improvements. The concept was first developed by Hiroyuki Hirano and it was further developed by Takashi Osada around 1980's which was commonly adapted by Japanese firms to enhance human capabilities while improving the productivity [142], [143]. 5S has five words begin with "S" where it got its name and they are Seiri, Seiton, Seiso, Seiketsu and Shitsuke, in Japanese language which means Sort, Set in order, Shine, Standardize and Sustain in English. The five words can be explains as follows in literature.

Seiri – Sort: it is the first step of 5S and stress to remove all unnecessary or surplus objects from the workplace that has no immediate requirement for ongoing operations [144].

Seiton – Set in Order: The second word which requests to keep all the sorted items in right places where they are frequently required for the smoothness of operation. The users must be motivated to place objects in right places where it belongs or at its point of use which help improve the visual management of work place [145].

Seiso – Shine: After the removal of unnecessary items away from the work station while re-placing the necessary objects in right places according to the utility, it was necessary to set the sanitizing or cleaning standards [146] where a cross-functional team should decide on the required cleaning standards [147] for the operation.

Seiketsu – Standardize: The maintenance of the work place was mandatory after organizing and cleaning of the place, thus standardizing is required [148] for the continuation of previous achievements [147].

Shitsuke – Sustain: The success of a 5S implementation is basically depends on the sustainability of the program where benefits of above 4S were easily measurable and visually observed. But without self-discipline which was the element of sustainability, can be momentary and transform back to the initial messy workstation [149].

According Goetsch and Davis [150], continuous cost reduction and improvements of quality were critical requirements for any organization to stay on business in any competitive marketplace. Since 5S was very popular around the world today as a total quality management tool (TQM), it has absorbed into many industries as well as service organizations which was further associated with other major TQM tools such as Kaizen, Continuous Improvement, Six Sigma, Just-in-Time (JIT) [151] as well as Lean Management. Thus 5S was helping many industries to reduce the waste while optimizing the productivity through maintaining an orderly workplace with the use of visual cues to accomplish highly consistent operational outputs

[152], [153], [154]. Nevertheless, some of the empirical analysis had demonstrated that the successful implementation of 5S significantly improve the organization's financial and operational performance [155].

The Sri Lankan tea industry has adapted various quality tools to improve productivity, efficiency, customer satisfaction, food safety as well as environmental sustainability and social wellbeing. However, these systems were operated separately where it generated additional costs to the operation while making it is complex as well as not properly operated according to the requirements. Due to that fact, most of the adapted systems were abandoned, weakly operated or virtually existed only for auditing dates. In contrast, 5S implementations which were initially promoted by the Sri Lanka Tea Board had gained the significance where it was drifted in to the all existing tea factories up to a certain extent while some of the factories were operating at extraordinary conditions. Most of the employees were also aware about 5S and they practiced it in their workplaces. The research was further intended to study and select the basic areas of interventions compatible with ISO 22000:2005; which could be merged to ISO 22000 FSMS with or without modifications to its original implementation objectives while designing relevant document formats.

Traceability in Sri Lankan Orthodox Black Tea Manufacturing Process

Considering current trends in export market, tea is moving worldwide as a healthy beverage which has advantages of availability but disadvantage of complex supply chains. Today global food competition is more intense where stakeholders started to adapt their mind sets toward a more holistic approach on supply chain management while focusing on food safety and traceability in a farm-to-fork perspective [156] because food industry has drastically changed during recent decades [157]. Nevertheless, many governments are improving food safety measures to safeguard their citizens by increasing control at all stages of food production, processing and distribution with hazard analysis critical control point (HACCP) and traceability based food safety management systems (FSMS) such as ISO 22000: 2005. According to Food and Agriculture Organization of the United Nations, traceability requirements are gaining significance in the global tea economy, with the potential for significant impacts on the world's tea producers, which is expected that the demand for traceability will result in widespread industry restructuring, which will require institutional responses within each of the major tea-producing regions [158]. The traceability is used as a tool to track product movement through manufacturing and distribution chain up to end user and backward to raw material supplier. Today consumer safety is the firm concern, which is a matter of cooperation between all actors involved in food supply chain rather than a confrontation; with the sharing of information, use of common standards and languages to pinpoint the food safety issues [159].

Concern over food safety is becoming central to supply chain restructuring in tea. Due to the costs, sampling and methodological constraints associated with the monitoring of finished food products, there is a trend towards preventing contamination at source through monitoring of estate and factory processes. This is associated with traceability based food safety management systems such as Good Agricultural Practices (GAP), Good Manufacturing Practices (GMP), ISO 22000 and HACCP certification. The second driver of traceability is consumer demands for independent verification that teas have been produced in the absence of abusive labour practices, by respecting worker rights and providing a living wage to smallholders, and by not polluting the environment or threatening biodiversity [158].

According to the ISO 9001:2000 standard, traceability is defined as "The ability to trace the history, application or location of product or service that is under consideration"[160], which is a process that makes it possible to find the traces of the various steps (manufacturing steps, sources of its raw materials with relevant suppliers, controls and tests carried out) and locations (storage locations, equipments used to manufacture or handle it, direct customers and end users) a product has passed through from its creation through to its final disposal [159]. Traceability requirements in food supply chain due to consumer concerns were started to rise in the later part of nineteenth century with the discoveries of microbes as well as vitamins which were further increased due to the development of supply chains followed by the improved production methods [161]. The product traceability is the ability to follow the food movement through specified stages of production, processing and distribution in one step forward and backward at any given place of the food supply chain, while facilitating efficient product recall in the event of hazard left the production facility. As a result of global, integrated partner oriented approach throughout food supply chain, every operator must identify their product in a unique way while recording their destinations with the links between incoming and outgoing products as well as end users on databases [159].

Further, product traceability helps to determine the origin of a food safety problem and to comply with legal requirements while meeting consumers' expectations for the safety and quality of purchased products [162]. The aim of all these measures is to safeguard consumers from biological, chemical and physical hazards that may be present in food while implementing full traceability in food supply chains.

According to the Kelepouris et al. [163], Morrison [164], Van Dorp [165] Viaene & Verbeke [166] full traceability or traceability throughout entire supply chain is essential for ensuring food safety and quality which was mainstream requirement to regain or maintain consumer confidence, where traceability cannot improve food safety or quality itself although it can provide necessary information and keep track of product movements [167].

Considering the complex nature of tea supply chain as well as the complex manufacturing operations involved in the tea industry, it is very difficult to identify the movement of a given supplier's raw material throughout the production process at a single glance. It was rarely discussed subject in many researches, where black tea's health benefits, chemical structures, disease prevention etc., has been extensively studied by many researchers. The existing manufacturing processes are also very old when compared to most of the modern food industries where traceability was not a major concern over a century. Hence the propose of this study was to find out the major drawbacks in the area of traceability by analyzing factors affecting the traceability process in orthodox black tea manufacturing and to find out possible solutions to the issues identified in the study.

As to the results, traceability was found in place up to a certain extent in all the tea manufacturing processes from one step forward. It was mostly limited to the made tea up to auctioneer or the wholesale buyer after completion of the manufacturing process, where made tea could be traced from manufacturer through product coding, manufacturing date, brand, and name of the manufacturer up to auction and through auction records purchaser can be located.

According to the CBI, the consumer markets were dominated by the popular blended brands which can contain up to 36 different types of tea varieties that were blended in the consuming country to ensure the unique taste of their brand at a competitive price where tea buyers source different teas from around the world mainly China, India, Kenya and Sri Lanka [40]. Accordingly, buyers blend tea after purchasing according to their brand's requirements depending on the market and cost. Therefore, the process was complex and the traceability will lead to a bunch of manufacturers rather than a single manufacturer.

Thus rest of the traceability was not studied because most of the purchases were export oriented and it was beyond the research objectives. However, that can be traced up to the consumer since no processing was involved other than blending, repacking or value addition according to the consumer preferences.

Thus traceability of a product complains can be elaborate as consumer claim to retailer, who claim to wholesaler and who will then inform the buyer through auctioneer and the manufacturer. If there is a product recall, the same channel can be used vice versa.

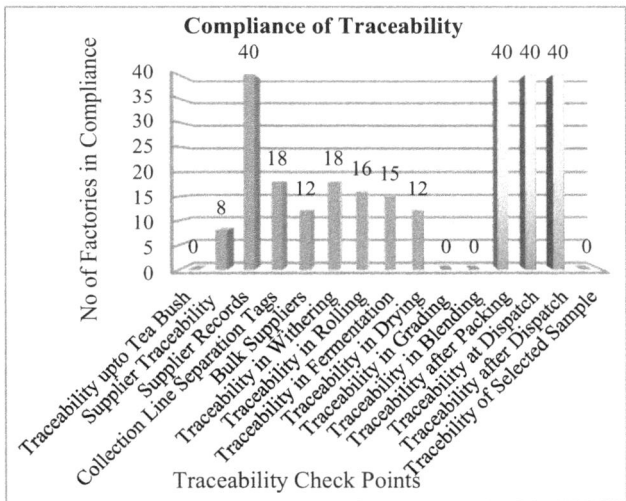

Figure 9.4: Constraints and compliances of traceability in low grown orthodox black tea supply chain

The results of the study was given in the bar chart above (figure 9.4), which illustrate the traceability practices as well as the achievements in the factory flow level. Further to the results obtained traceability up to tea bush, traceability in grading, traceability in blending and traceability of sample back to the supplier wasn't fully complying according the study sample ($p < 0.05$). On the other hand, supplier records, traceability after packing, traceability at dispatch and traceability after dispatch was fully traceable or all the tea factories were in compliance with these four factors (p = 1) . As a rule of thumb, other factors had varying degree of traceability compliances, which indicated that those areas were more or less neglected but it was not impossible. The figure 9.5 below shows the probability of implementation and it further prove that what were already implemented and what was not achieved due to explained reasons or neglected according to the representative sample.

When considering traceability in one step backward, none of the brought leaf based tea factories were able to exactly locate the farmer or the field where green leaves were harvested. Large estates were able to locate the field of the harvesting carried out, but they also unable to locate the records of exact tea bush or the labourer who harvested a given green leaves which mostly process in their own production facility. On the other hand, smallholder suppliers were more critical because their leaves were mixed each other while collecting and transport, where traceability was only possible up to a bunch of suppliers.

Bearing in mind the traceability up to tea bush which was the primary objective, a quantity of 4.50 – 4.65 kg of green leaves were required to manufacture 1kg on made tea (final product) [41]. 1kg of green leaves were to be harvested from 20 -

30 tea bushes with average yield where it is extremely difficult or almost impossible to practically locate the exact tea bush in the case of a product complain or hazard occurrence.

Figure 9.5: Implementation probability of traceability in low grown orthodox black tea supply chain

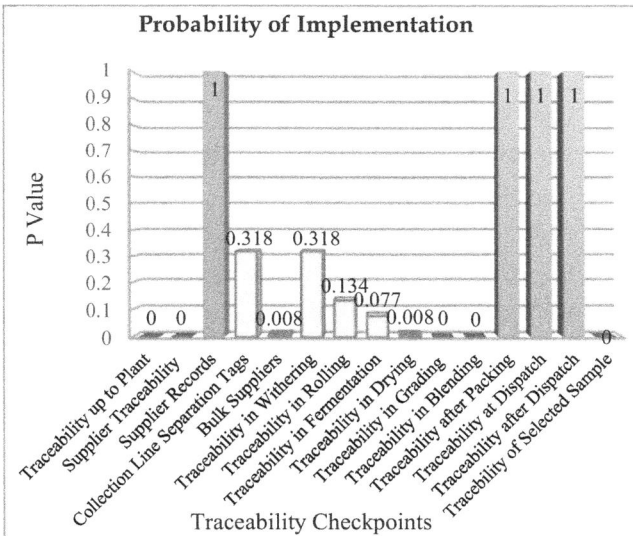

The most possible paper based system could reach the leaf collectors or bunch of suppliers approximately, but it was impossible to locate the exact supplier or tea bush who contributes to the issue. Thus one-step backward traceability in orthodox black tea manufacturing was rarely achieved due to following reasons in smallholder suppliers. Most of the suppliers were smallholders who had less than one hectare of lands, which produce less than a single roll charge in a one supply of green leaves. It was usually varying around 3 – 15 suppliers at a time.

Tea cultivation requires a relatively small investment, and the risk of complete crop failure is low. Problematic issues for smallholders producing bulk tea products include low farm gate prices, poor extension services and limited marketing channels [40]. As a result, their tea leaves were mostly collected through leaf collectors. According to keleporis et al, successful implementation of supply chain traceability calls for co-operation among all the stakeholders of the supply chain [163]. According to the interviews conducted, leaf collectors were not educated enough by manufacturers to keep traceability records. On the other hand, cost of additional labour and transport was high, where leaf collectors were operating at little margins. Nevertheless, leaf collectors as well as factory employees also believe that tea leaves are a uniform and homogenous raw material; origin in similar fields which cannot be distinguished each other very easily. Thus it was very difficult to motivate them to practice traceability in supply chain, where extreme competition for leaves in the market has the bargaining power for them to change the manufacturer at any time without notice.

After withering during release to the rolling process, small quantities were bulked or large quantities segregate to make a single batch of size about 250kg depending on the dryer capacity, which was call the charge, [29] and this activity increased the mixing of direct smallholder suppliers as well as leaf collectors. According to the study sample (as shown in the figure 9.4 – traceability compliance), 18 factories maintain collector bulks with identification tags ($p > 0.05$), thus the leaf could be traced with time records backward to leaf collectors as well as bunch of smallholder suppliers at a given date but it was impossible to trace up to the exact supplier or the tea bush. The same leaves could be traced forward through rolling, fermentation, drying up to fiber removal, easily with time records when the production was arranged in such a way.

Typically, orthodox black tea manufacturing process was much more complicated unlike other production processes (e.g. CTC) due to the different separation techniques employed as well as the number of different varieties produced. Hence, most critical part of the tea traceability systems was to trace the product within the manufacturing process. The sifting/grading was the key to number of different tea varieties and ungraded dry tea leaves initially pass through Myddleton after fiber removal.

According to Samaraweera [29], the Myddleton bubble tray stalk extractor has an oscillation motion of 200 to 220 oscillations per minute of two aluminum bubble trays with perforations where bubbles impart a string action at the bottom of the layer which makes long particles such as fibers or stalk to float. It helps small and round particles to settle at the bottom of the layer while top particles to fall through perforations. To achieve this action, it must be feed with a continuous thick layer without exposing the bubbles, which will sort the stalky long leaf portion by passing through a tray that is used to separate long leaf particles and stalks at the beginning of the sorting process [29].

Thereafter, Chota and Michie shifters were used to separate uniform sizes and then to different varieties. Michie oscillatory shifters were equipped with one or two trays stacked one on top the other with a slight slopes towards forward direction that gives oscillatory motion on the vertical plane which help particles to jump forward at 250 to 260 oscillations per minute. Michichi shifter was used to separate long and wiry particles which were basically used to produce long leaf and wiry grades [29]. After separations, semi graded tea was then passed through a winnower to separate according to the weight. The winnower is used to separate light tea particles such as flakes, fibers, tea fluff and extraneous matters such as sand or dirt [29].

However, this process was very complicated and very difficult to understand easily for an average person. Considering 1st dhool, 2nd & 3rd dhool as well as the big bulk, given generic process flowcharts (figure 9.6, 9.7 and 9.8 below) were drawn where it was obvious that single tea leaf could pass many paths before it end up in a specific product due to the weight, size and the shape of the leaf of a shoot based on the way it was rolled in the orthodox rollers.

On the other hand, consider the same leaf if it broken to small sizes and large sizes they normally end up in two different grades. The same shoot's tender leaves will mostly end up in first grades while matured low weight leaves will mostly end up in second grades based on the weights. Considering the complexities involved, the figures 9.6, 9.7, and 9.8 shows the number of different grades produced in a given dhool. It was almost 32 for a first dhool (figure 9.6) considering the generic processes without 2nd or 3rd grade products where their sizes were different from one another as well as different parts of the tea shoot.

This phenomenon was almost same for all the dhools where it wasn't a very easy process to keep records of the product movement. Nevertheless, the second grades were much less discussed in the orthodox black tea manufacturing process but those amounts were also massive and their movements were never recorded in the industry on accurate manner, where the existing traceability practices were not complete. The existing systems had introduced various documents to keep traceability tracks on the product movements, but none of them were able to comply with the theoretical traceability requirements according to the definition of traceability [160].

According to the given process flow diagrams below, one tea leaf might pass many times through Myddleton, Michie and Winnower as well as the Colour Sorter before it reached the final grade. Some leaves may recycle the same path again and again with mixing new coming ungraded teas such as throughput of 5mm Myddlton which go back again to Chota (red line – figure 9.6, 9.7 and 9.8) and passing through same sieves before end up in a final grade. On the other hand, each time the product went through winnower, based on the weight; it separates the product in to 1st grade, 2nd grade, 3rd grade and fiber while removing stones and other extraneous particles and reducing the bulk into small quantities. 1st grade was proceed with next steps of manufacturing process, while 2nd and 3rd grades were followed separate pathways which generates second quality tea products which were usually called second selling mark products. Considering 1st dhool, 2nd dhool, 3rd dhool and big bulk, when each of these process flowcharts were equally complicated and largely distributed, it was very difficult to follow a paper based or even a digital method to track down the traceability of a specific tea leaf.

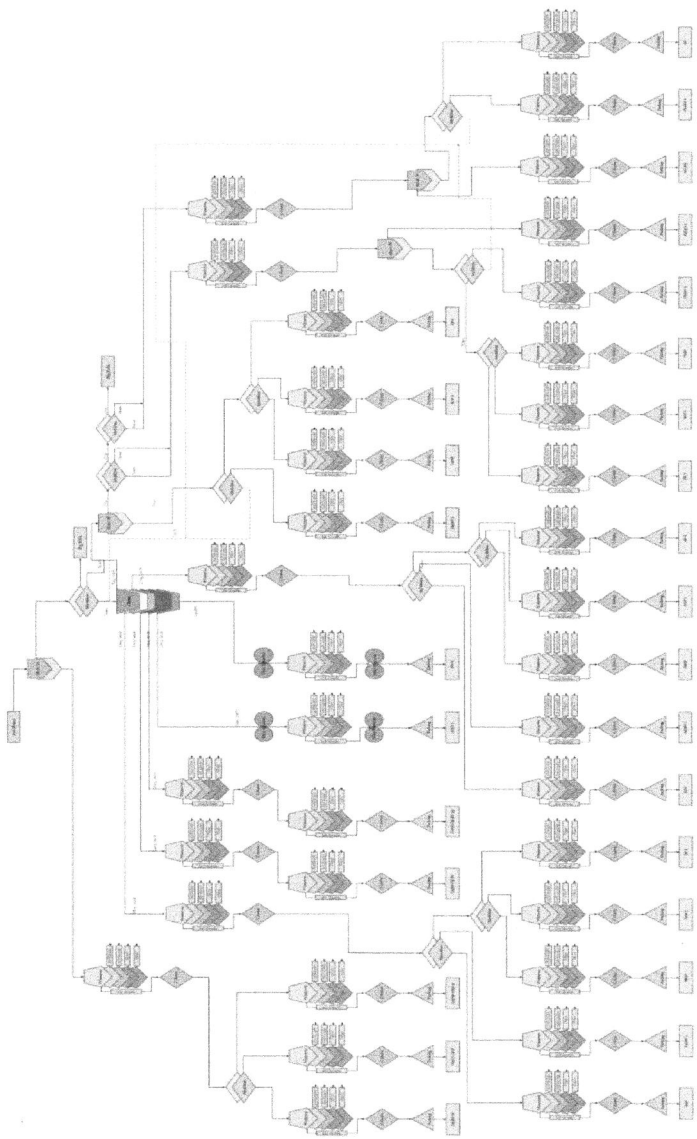

Figure 9.6: Graphical representation of generic process flow chart for the 1st Dhool sifting program of orthodox black tea

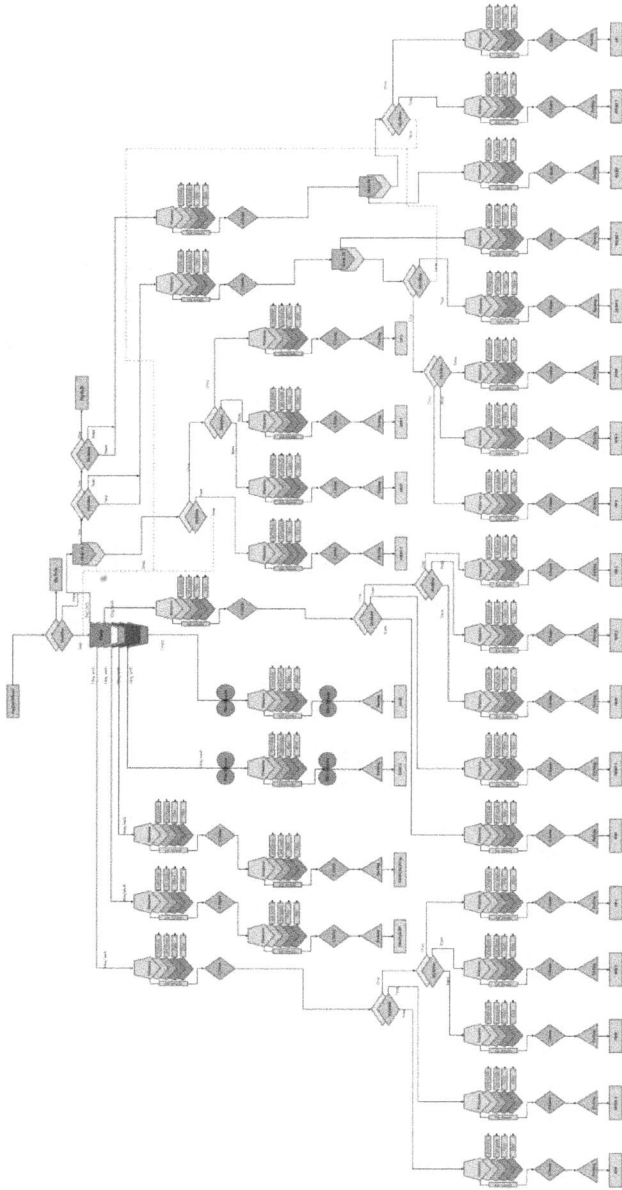

Figure 9.7: Graphical representation of generic process flow chart for the 2nd & 3rd Dhool sifting program of orthodox black tea

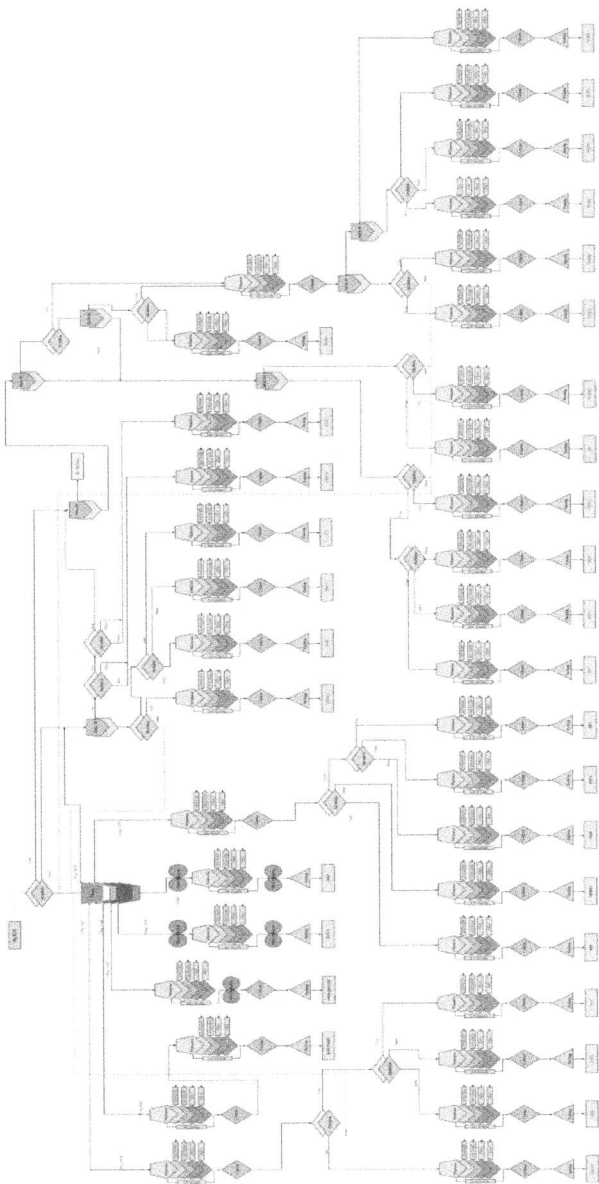

Figure 9.8: Graphical representation of generic process flow chart for the Big Bulk sifting program of orthodox black tea

Due to the continuous feeding and requirement of uniform thickness in Middleton shifter; batches couldn't be separated accurately rather than guessing. Thus there must be number of documents and complex storing systems to identify different batches as well as batch wise separation techniques which will add more cost to the production. While dried tea was passing through shifters within grading room operations, the same bulk was separated into small different grades based on leaf size, type, dhool, weight and buyer requirements where orthodox black tea has around 30 different grades basically which further creates issues in reaching the initial leaf supplier.

There are various other processes involved, but these processes were the major critical points for the traceability issues in orthodox black tea manufacturing unlike the cut, tear and curl (CTC) process where it is easily traceable due to non-complex process which give less varieties as well as large quantities. In orthodox process, the output of a single grade was small where several batches were required to prepare a standard quantity for packing or as to buyer requirements, thus bulking is needed.

The blending (figure 9.9) was carried out to give specific product characteristics to a given tea product where different grades are mixed together. Hence this process further increase the traceability issues while increasing the number of raw material suppliers as well as individual tea plants contributing to a given tea product.

Nevertheless, these areas still can be traceable with a proper organization of production flow and better recording systems which was not available in any of the systems evaluated. The cost is one of the major factors which decide the implementation of such measures, because even though the factories are fully dedicated and implemented such systems, the buyers do not pay attention to such efforts while they were involved in purchasing. Thus it is inappropriate to request complete treatability from manufacturers without providing solutions to the areas mentioned in the study.

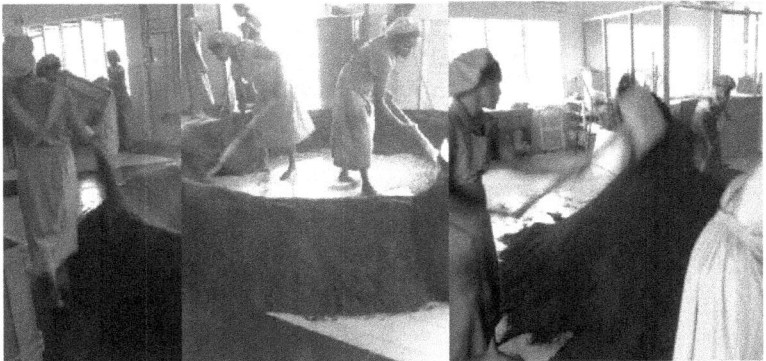

Figure 9.9: A Typical blending operation

On the other hand, if orthodox black tea manufacturing was considered as a bulk material, the traceability can be achieved with less hazardous work load where bunch of suppliers or fields were the primary objective. Then paper based traceability systems can be developed to trace single day's production which can be more easily segregated while storing and recording where it can be traced through grading and blending up to packing. Nevertheless, it can be trace one step backward practically up to bunch of suppliers through existing records while adding few new records within grading operation from drier out to finished product packing. Rest of the areas of manufacturing namely receiving, withering, rolling, fermenting and drying can be managed through addition of time records or codes to the existing food safety management systems.

Finally, if the cost is bearable to the consumer, DNA barcoding is a widely used molecular-based system, which can identify biological specimens, and is used for the identification of both raw materials and processed food, which can be possible to employ to identify exact issue concerning samples [168]. On the other hand, Radio Frequency Identification (RFID) tags also can be integrated to track down the raw material movements as well as finish product by using the passive RFID tags with the bulk storing bins from field to the factory and then after drier moth to the packing since raw material and finish product both moved within plastic crate which can be continually used with a reader. The passive RFID tags are designed to depend on the tag reader as its power source where reader unit can communicate with the tag within 6 meters which are low in product cost and it is manufactured for disposal with disposable consumer products [169]. Considering current developments in RIFD tags technology, it possibly can be used in the complicated operations in orthodox back tea manufacturing process to track down intermediate product movement within the factory environment where the production should plan according to the requirements while synergizing with existing paper based systems to make it workable. However given model will be able to identify the traceability on the bulk movement where individual supplier and the end user can be located. But these technologies are not currently applied in tea trade where it will not guarantee the accurate identification of exact tea bush in an event of hazard occurrence which needs further research to explore the possibilities of exact application of the given technologies.

Accordingly, the research recommends considering tea as a bulk material like edible oils in the event of processing where it is really practical to consider traceability after made tea left the factory and if there is any food safety issue manufacturer has to address all the suppliers in the supply chain and check their plantations for identification of abnormalities, if necessary. If the made tea is considered as a bulk material, it will be helpful to develop user friendly, realistic and practical paper based traceability models which will not increase the product cost or unnecessary documentation works for the quality assurance system operators.

Common Minor Concerns in tea trade which needs no financial requirements or negligible financial requirements to implement by the tea factory management.

1. Prepare well defined functional organizational chart and assign in writing respective responsibilities according to it. The responsibilities should be in written documents.
2. Establish proper writing, approving and distributing written procedures for appropriate persons.
3. Activate one or more individuals with written responsibility for supervision of plant sanitation program.
4. Perform complete inspection of the entire plant's food safety at least once a month.
5. Implement and improve the functions of the multidisciplinary food safety team, assign responsibilities and introduce records and other documentation.
6. Monitor movements of employees who are continually moving through place to place to minimize cross contamination.
7. Appropriate monitoring procedures for hand washing while entering to the production area should be implemented.
8. Maintain appropriate documents for visitors who enter the production area and prepare a method to walk through clean area to dirty areas.

9. Special provision must be made to interrupt the production for cleaning and disinfecting at least every eight hours.
10. Prepare appropriate cleaning plans and schedules with maintenance and inspection records.
11. Written, formalized cleaning procedures must be implemented and practiced for every department.
12. Assign an operative to maintain cleaning schedules and adequate documented procedures.
13. The cleaning schedules must dictate the frequency and method of cleaning and disinfecting agents that are to be used for all plant, equipment and surroundings.
14. All cleaning and disinfection agents must be stored safely in a designated, secure area, off the production floor. They must only be used in accordance with the manufacturer's instructions and their use kept to an absolute minimum when production is in progress.
15. The responsibility for dispensing cleaning and disinfection agents must be with designated personnel and their usage monitored to ensure that the correct, safe percentage dilution is being used.
16. Containers and equipment for the preparation and dispensing of sanitizing agents must be kept under the control of designated persons.

17. An up-to-date list (Check list) must be available for all cleaning and disinfection agents used on the premises and all operatives and first aiders aware of their chemical content in the event of an accident.
18. There should be procedures, check lists and records available for clean as you go and analysis of chemicals been used.
19. Improve good housekeeping and cleaning.
20. Improve cleaning of site and removal of pest/rodent harbourage sites as much as possible. Remove all unwanted and redundant equipment or parts from the plant.
21. Adequate cleaning regimes with records should be scheduled and monitor to enhance the hygienic conditions of the floors, walls, windows, window sills, lights and ceilings.
22. Adequate records must be maintained for the maintenance of compressed air supplies.
23. Labeling system should be adequate and should be in logical order or other method to ease the withdrawal.
24. Rejected items or items pending for approval should be stored in separate areas.
25. There should be a documented procedure for handling of defective batches or products.
26. Implement procedures and records for handling damages and accidents during storage and distribution.

27. Implement a system to internally inspect vehicles prior loading.
28. Maintain adequate procedures for inspecting, monitoring, recording, loading, transit and unloading.
29. There should be adequate and continuous training schedules for operatives to improve the plant's sanitary requirements for continual improvement and update.
30. Their trainings must be planned according to the assigned tasks and workplace.
31. Pest control program should be planned with relevant guide lines for frequent audits and specific works.
32. Internal bait stations and places should be planed according to the requirements.
33. Adapt to a proper documentation of pest control system.
34. A member of management team should accompany with the pest control officer and the recommendations should be implemented within the given time.
35. Removal of resident animals from site and main gate should be closed at all the time.
36. Documents for safety and application of approved baits and pesticides must be available.
37. Hygienic conditions of rest rooms must be improved with good training for operatives.
38. Toilets should be cleaned and well kept.

39. Hands should be thoroughly dried, nails kept short and well-manicured and cleaned by using a nailbrush. In addition to that, hands must be washed when immediately before putting on protective clothing, particularly important in the case of high risk, personnel, immediately before commencing work or entering production areas, after handling debris, refuse or food waste, if they become soiled or visibly contaminated, after visiting the toilet and after blowing the nose or touching the mouth.

40. Behaviours of the operatives must be controlled at appropriate level.

41. Implement a system to monitor visitors in to the plant.

42. Keep adequate documents against temperature control and monitoring.

43. Adequate documentations should be implemented according to the changes in the plant.

Common Major Concerns are the events that management needs considerable financial investments to improve the food safety and quality of the product manufactured. The following list may help to improve the tea industry.

1. Management has to provide necessary capital requirements for the refurbishments that are necessary in both withering and production areas.
2. Prepare up-to-date design and layout plans.
3. Provide appropriate equipment to clean and rearrange the stores according to the production requirements and separate areas for specific materials as well as packing materials.
4. A system should be established to monitor entering to production area by different employees such as use of different coloured uniforms with restrictions.
5. Conduct testing and maintain adequate records and documents on portability of water.
6. Protective goggles, gloves gumboots and overalls must be used by all persons handling cleaning chemicals as instructed on the manufacturer's guide on site and their use regularly enforced by management.
7. There should be separate facilities for equipment and utensil washing, adequate storage space for complete segregation of dirty and clean utensils which can be

physically separated from production area with adequate facilities for drying and storing hygienically.

8. Take appropriate measures to minimize the inadequacy of machinery and equipment sitting where applicable and possible.

9. Where appropriate, non-perfumed barrier creams or alcohol based skin sanitizers should be provided.

10. Construct facilities for hand washing, showering, toilet facilities and dedicated changing area for workers.

11. Adequate boot washing facilities should be provided for entrances, where risk of cross contamination is possible.

12. Waste bins must be covered with a pest proof lid at its top to prevent pest harbourage and rodent attraction.

13. Use colour coded or different containers for waste collections with clearly identified for specific usage.

14. Worker rest rooms should be improved with appropriate space and clean environments.

15. The space and facilities for factory debris should be improved.

16. Employ a Pest control officer with relevant criteria mention in job description.

17. Implement a system for safe and hygienic disposal of pests.

18. Prepare accurate internal bait satiations and system by the pest control officer.

19. Adequate measures should be taken to improve the number of traps been used.

20. Use appropriate covering for entry and exit points where necessary to prevent entering rodents and pests. Use plastic strip curtains where necessary according to the specific requirements.

21. There should be adequate and continuous training schedules for operatives to improve the plant's sanitary requirements for continual improvement and update. Their trainings must be planned according to the assigned tasks and workplace.

22. All ancillary and production areas should be covered with pest control plans.

23. Exit and entry points of drainage systems should cover with appropriate netting. Open drains should be covered with grill or a mesh.

24. The floor has to be sloped towards the drains when and where required.

25. Grease used in equipment and machinery should be in food grade.

26. Replace the wooden scraperboards with stainless steels.

27. Take appropriate actions to minimize damage to the floor and repair damaged places as quickly as possible with appropriate flooring materials where floor requires specific attention.

28. Use food grade paints for all production areas.
29. Seal all holes and remove rough surfaces in the walls or ceilings with appropriate methods. Repairs should be adequate and compatible with original finish.
30. Cover all the windows and natural lightings with plastic nets when and where it is necessary.
31. Employ humidity controllers to maintain optimum conditions for fermentation.
32. Train the drivers with relevant procedures and GMP requirements.
33. Implement adequate procedures to minimize the lower graded brought leaf mixing with high quality leaves and product damage during transport.
34. There should be separate facilities for equipment and utensil washing, adequate storage space for complete segregation of dirty and clean utensils which can be physically separated from production area with adequate facilities for drying and storing hygienically.
35. Improve cleaning of site and removal of pest/rodent harbourage sites as much as possible. Remove all unwanted and redundant equipment or parts from the plant.

36. The exterior of the factory should be cleaned and hygienic with good maintenance of factory buildings.
37. Complete physical separation for damaged, returned or infested products.
38. Adequate monitoring, maintenance of waste disposal system and waste storage facilities should be improved.
39. All the sinks fixed inside and toilet should be hand free operated.
40. Adequate drying facilities must be provided.
41. Monitor adequate hand swabs at frequent times with random employees.

References

1. Berggreen, S.C., 2014, *Deep Drink: Notes on Mediating Tea in Public Imagination*, International Journal of Humanities and Social Science.

2. Chow, K. and Kramer, L., 1990, *All the tea in China.* Accessed through http://www.chinabooks.com/ Excerpt/alltea.html

3. Wright, L.P., 2005, *Biochemical analysis for identification of quality in black tea (Camellia sinensis)*, PhD thesis, University of Pretoria, South Africa.

4. Tea - A Brief History of the Nation's Favourite Beverage, 2015, UK Tea and Infusions Association, Accessed through http://www.tea.co.uk/tea-a-brief-history on 14/07/2015.

5. A Fair-trade Foundation Briefing Paper, 2010, Stirring up the tea trade, Can we build a better future for tea producers, Accessed through: http://www.beverage standardsassociation.co.uk/PDF/FTrade_Stirring_ up_the_tea_trade(Jan10).pdf, Accessed through: http://www.iso.org/iso/iso-22000_food_safety.pd f.

6. Willson, K.C., 1999. Coffee, Cocoa and Tea. CAB International, Wallingford, UK.

7. Anonymous, 2002. The Tea Growers Hand Book. 5th Edn. The Tea Research Foundation of Kenya Printing Services.

8. Elliott, E. C. and Whitehead F.J., 1928, *Tea planting in Ceylon*.

9. Munasinghe M., Deraniyagala Y and Dasanayake N., *SCI Research Study Report: Economic, Social and Environmental Impacts and Overall Sustainability of the Tea Manufacturing Industry in Sri Lanka*, 2013. Accessed through: http://www.sci.manchester.ac.

uk/sites/default/files/Economic,%20Social%20an
d%20Environmental%20Impacts%20and%20Overal
l%20Sustainability%20of%20the%20Tea%20Manufa
cturing%20Industry%20in%20Sri%20Lanka_0.pdf

10. TRI, 2003, Anonymous.

11. http://www.toptenofcity.com/commerce/top-10-largest-tea-producing-countries-2011.html

12. Department of Census and Statistics, 2005 Census of Tea Small Holdings, Sri Lanka.

13. Department of Census and Statistics, 2012.

14. Lokunarangodage, C.V.K, Wickramasinghe, I. and Ranaweera, K.K.D.S., 2012, *Development of a Profitable Business Model to Promote Consultancy on Food Safety through Evaluating Current Practices in Tea Industry*, Proceedings of International Conference on Business Management. University of Sri Jayewardenepura.

15. Mohamed, Z. M. T. and Zoysa, A. K. N. (2004). Current Trends and Future Challenges in Tea Research in Sri Lanka. Proceedings of the symposium on plantation crop research, Tea Research Institute of Sri Lanka.

16. Scott, R., 2009, *Sri Lanka Aims to Lure Investors*, www.bloomberg.com, August 26. Accessed January 24, 2010.

17. Mohammed, M. T. Z. and Zoyza, A. K. N., 2008, Chapter 02: An Overview of Tea Industry in Sri Lanka, *Hand Book on Tea*, Tea Research Institute Sri Lanka., Page 04 – 09.

18. FAO: 2012.Current Situation and Medium Term Outlook for Tea. Committee on Commodity Problems, FAO, 20th Session of the Intergovernmental Group on Tea. 30th January – 1st February 2012, Colombo, Sri Lanka.

19. Grigg, D., 2003, The worlds of tea and coffee: Patterns of consumption, *Geo-Journal*, Volume 57, No. 4, p. 283-294.

20. Products/Black Tea, 2008, Available at http://www.irvingtea.com/en/products/Black-Tea/ Accessed April 07, 2010

21. Ute Williges, 2004, Status of organic agriculture in Sri Lanka with special emphasis on tea production systems (Camellia sinensis (L.) O. Kuntze), PhD Thesis, Faculty of Plant Production, Justus-Liebig-University of Giessen, Accessed through: http://geb.uni-giessen.de/geb/ volltexte/2005/2315/pdf/WilligesUte-2005-02-10.pdf

22. Franke, 1994. Nutzpflanzen der Topen und Subtropen, Baud 3, Ulmer Verlag, Stuttgart, BRD.

23. George, E.F. and Sherrington, P.O., 1984, *Plant propagation by tissue culture. Hand book and directory of commercial laboratories.* Exegetic Limited, London.

24. Roberts, G. R., 2008, *Hand Book on Tea*, Chapter 19: Principles of Tea Manufacture, Tea Research Institute Sri Lanka, p. 261 – 264.

25. Sen, R. N.; Gangulgi, A. K.; Ray, G. G.; De, A; Chakrabarti, D., 1983, Tea-leaf plucking-workloads and environmental studies, Ergonomics 26: pp887–893.

26. Wijeratne M. A., 2008, Chapter 9: Harvesting of tea, *Hand book on tea*, Revised by Wijeratne M. A., Tea Research Institute Sri Lanka, pp. 94-104

27. Owour, P. O. and J. E. Orchard, 1989, Changes in the biochemical constituents of green leaf and black tea to withering: a review. Tea 10(1): 53 – 59.

28. Das, S. K., 2006, Further Increasing the Capacity of Tea Leaf Withering Troughs, *Agricultural*

Engineering International: the CIGR E-journal. Manuscript FP 05 012. Vol. VIII.

29. Samaraweera D. S. A., Chapter 20: Technology of Tea Processing: *Hand Book on Tea*, Revised by Samaraweera D. S. A. and Mohammed, M. T. Z., Page 265 – 322. Tea Research Institute Sri Lanka., 2008.

30. Guang L., 2007, Ling C. C., ed., The Traditional Processing of Wuyi Rock Teas: An Interview with Master Ling Ping Xang, *The Art of Tea* (Wushing Book Publisher), (2): 76-83

31. Roberts, E. A. H. (1958), "The Chemistry of Tea Manufacture", J. Sci. Food Agric. 9

32. http://www.tea.lk/tea-production/: Accessed on 19/04/2015.

33. Varnam, Alan H.; Sutherland, J. M. 1994, Beverages: Technology, Chemistry and Microbiology, Springer.

34. Chen (陳), Huantang(煥堂); Lin (林), Shiyu (世煜) 2008, *The first lesson in Taiwanese tea* (台灣茶第一堂課：頂尖茶人教你喝茶一定要知道的事！), 如果出版社, ISBN 978-986-6702-21-1

35. Mohamed, Z. M. T. and Dahanayake, D. L. H., 2003, Studies and opinions on up country tea manufacture. In: Twentieth century tea research in Sri Lanka (Ed.) Modder, W.W.D. Tea Research Institute of Sri Lanka. 361 – 364.

36. Bailey, R. G., Nursten, H. E., McDowell I., (1992). "Isolation and analysis of a polymeric thearubigin fraction from tea." Journal of the Science of Food and Agriculture 59: 365-375.

37. Bailey, R. G., Nursten H. E., and McDowell, I., (1993). "The chemical oxidation of catechins and other phenolics: a study of the formation of black tea

pigments." Journal of the Science of Food and Agriculture 63: 455-464.

38. Mauskar, J.M., 2007. Comprehensive Industry Document on Tea Processing Industry. Accessed Through:
http://cpcb.nic.in/upload/Publications/Publicatio n_496_Test% 20publication.pdf

39. Embole, S. L., 2011, *Tea for Dummies*, Black tea, pp. 84 – 89. Accessed through:https://sindromeembole. files.wordpress.com/2011/04/tea-for-dummies.pdf

40. CBI Market Channels and Segments for Tea, '*Your trade route through the European market*', 2013, accessed through http://www.cbi.eu/system/ files/marketintel_platforms/2013_market_channel s_and_segments_tea_-_coffee_ tea_and_cocoa.pdf.

41. SustainabiliTea - Report on Sri Lankan Tea Industry., January, 2008. Institute of Social Development, Kandy, Sri Lanka. Accessed Through: www.somo.nl/ publications-en/Publication_3095/ at_download/fullfile.

42. Fuchs, H.J., 1989, Tea environments and yield in Sri Lanka, Tropical Agriculture 5, Margraf Scientific Publishers, Weikersheim, Germany.

43. Schumann, G.L., American Pathological Association, "Why Europeans Drink Tea" from the book of Plant Diseases: Their Biology and Social Impact, Accessed through: http://www.apsnet.org /publications/
apsnetfeatures/Pages/ICPP98CoffeeRust.aspx

44. Boyle, R., 2012, Loolecondera, The Birth Place of Ceylon Tea. Accessed on 7th July, 2015, through: http:// www.serendib.btoptions.lk/article.php?iss ue=30&id=764.

45. Humbel, R., 1991, *Tea Area Changes in Sri Lanka:* Analysis of regional distribution, processes, mechanisms and correlating factors of changes in area cultivated with tea since 1956. Dissertation University of Zurich, Switzerland.

46. Sri Lanka Tea Board, 1996, *Tea Land Survey of Tea Smallholdings and State Owned Estates in Sri Lanka* 1994 -95, Tea Commissioner's Division, Sri Lanka Tea Board, Colombo, Sri Lanka.

47. Bets, J., 1993, Agrarische Rohstoffe und Eutwicklung, Teewirtschaft und Teepolitik in Sri Lanka, Indien und Kenia. Schriften des Deustchen Ubersee–Instituts Humburg, Nr. 21, BRD.

48. Statistical Information on Plantation Crops – 2012; Ministry of Plantation Industries, Colombo, 2013, p 1 - 68.

49. Anon, 2003. Agricultural Profile of the Corporate Tea Sector, Tea Research Institute of Sri Lanka, 2003

50. Mohamed, M. T. Z., Galahitiyawa, G., and Chandradasa, P. B., 2003, Nett outturn of made tea to green leaf in Low-country. Sri Lanka Journal of Tea Science Vol 68, Part 1.

51. Janaka W., Suwendrani J., 2011, Implications of agri-food standards for Sri Lanka: Case studies of tea and fisheries export industries, Asia-Pacific Research and Training Network on Trade Working Paper Series, 104: 03 – 39.

52. www.referenceforbusiness.com/encyclopedia/Int-Jun/ ISO-9000.html. : Accessed on 17/04/2015.

53. Martincic, C. J., 1997, A Brief History of ISO - II (20/02/1997) - http:// www.sis.pitt.edu/~mbsclass /standards/martincic/isohistr.htm: Accessed on 17/04/2015.

54. The history of ISO 9000, 2012, www.british-assessment.co.uk/articles/the-history-of-iso-9000: Accessed on 20/04/2015.

55. Sroufe, R. and Curkovic, S., 2008, An examination of ISO 9000:2000 and supply chain quality assurance, *Journal of Operations Management*, Vol. 26 No 4, pp. 503- 520.

56. Sun, H., Li, S., Ho, K., Gersten, F., Hansen, P. and Frick, J., 2004, The trajectory of implementing ISO 9000 standards versus total quality management in Western Europe, *International Journal of Quality and Reliability Management*, Vol. 21 No 2, pp. 131- 153.

57. Van der Wiele, T., Van Iwaarden J., Williams, R. and Dale, B., 2005, Perceptions about the ISO 9000 (2000) quality system standard revision and its value: the Dutch experience, *International Journal of Quality and Reliability Management*, Vol. 22 No 2, pp. 101-119.

58. Aggelogiannopoulos, D., Drossinos, H. and Athanasopoulos, P., 2007, Implementation of a quality management system according to the ISO 9000 family in a Greek small-sized winery: A case study, *Food Control*, Vol. 18 No 9, pp. 1077–1085.

59. Bolton, A., 1997, Quality Management Systems for the food Industry: A guide to ISO 9001/2. London: Blackie.

60. Tufan K., 2007, The impact of ISO 9000 quality management systems on manufacturing, *Journal of Materials Processing Technology*, Vol. 186, pp. 207–213.

61. Webster's Ninth New Collage Dictionary., Safety: Accessed through http://www.merriam-webster.com/dictionary/safety.

62. FAO/WHO, 1997, Codex Alimentarius Guidelines for the Design, Operation, Assessment and

Accreditation of Food Import and Export Inspection and Certification Systems (CAC/GL 26-1997). FAO/WHO, Rome.

63. Motarjemi, Y., Mortimore, S., 2005, Industry's need and expectations to meet food safety, 5th International meeting: Noordwijk food safety and HACCP forum 9-10 December 2002. Food Control 16(6):523-529.

64. Burros, M., 1997 (24 January), Clinton to battle food borne illness. New York Times.

65. Garvin, D., 1993, Building a learning organization, *Harvard Business Review*, vol.71, July-August, 78-91.

66. Misterek, S.A., Anderson, J.C., & Dooley, K.J.,1990, The strategic nature of process quality, The Decision Sciences Institute, New Orleans, LA.

67. Trienekens, J. and Zuurbier, P., 2007, Quality and safety standards in the food industry, developments and challenges, *International Journal of Production Economics*, Vol. 113 No 1, pp. 107-122.

68. Karipidis, P., Athanassiadis, K., Aggelopoulos, S., and Giompliakis, E. (2009), "Factors affecting the adoption of quality assurance systems in small food enterprises", Food Control, Vol. 20 No 2, pp. 93-98.

69. Newman, E J., 2005, Accreditation, Quality Assurance/ Accreditation, Newman, Bucknell Associates, Wimborne, UK, pp. 485-489

70. Ziggers, G.W. and Trienekens, J., 1999, Quality assurance in food and agribusiness supply chains: Developing successful partnerships, *International Journal of Production Economics*, Vol. 60-61, pp. 271-279.

71. Luning, A. and Marcelis, J., 2006, A techno-managerial approach in food quality management

research, *Trends in Food Science and Technology*, Vol. 17 No3, pp. 378-385.

72. Manning, L. and Baines, R., 2004, Effective management of food safety and quality, *British Food Journal*, Vol. 106 No 8, pp. 598-606.

73. Mamalis, S., Kafetzopoulos, D. P., Aggelopoulos S., 2009, The New Food Safety Standard ISO 22000. Assessment, Comparison and Correlation with HACCP and ISO 9000:2000. The Practical Implementation in Victual Business: Paper prepared for presentation at the 113th EAAE Seminar "A resilient European food industry and food chain in a challenging world", Chania, Crete, Greece, date as in: September 3 - 6, 2009.

74. Meiselman, H., 2001, Criteria of food quality in different contexts, *Food Service Technology*, Vol. 1, pp. 67–84.

75. Juran, J. M., 1989, Juran on Leadership for Quality, The Free Press, Collier Macmillan, New York.

76. Olsen, J., Harmsen, H. and Friis, A., 2008, Linking quality goals and product development competences, *Food Quality and Preference*, Vol. 19 No 1, pp. 33-42.

77. Burlingame, B. and Pineiro, M., 2007, The essential balance: Risks and benefits in food safety and quality, *Journal of Food Composition and Analysis*, Vol. 20 No 2, pp. 139–146.

78. Holleran, E., Bredahl, M.E., Zaibet, L., 1999, Private incentives for adopting food safety and quality assurance, *Food Policy* Vol. 24, pp. 669–683.

79. Kontogeorgos A. and Semos A., 2008, Marketing Aspects of Quality Assurance Systems: The Organic Food Sector Case, *British Food Journal*, Volume 110, Issue 8.

80. Van der Spiegel, M., Luningy, P., Ziggers, G., and Jongen, W., 2004, Evaluation of Performance Measurement Instruments on Their Use for Food Quality Systems, *Critical Reviews in Food Science and Nutrition*, Vol. 44 No 4, pp. 501–512.

81. FAO, 1998, *Food Quality and Safety Systems* - A Training Manual on Food Hygiene and the Hazard Analysis and Critical Control Point (HACCP) System: Section 3 - The Hazard Analysis and Critical Control Point (HACCP) System. Publishing Management Group, FAO Information Division, Rome. Accessed through: http://www.fao.org/docrep/w8088e/w8088e05.htm

82. Khandke, S. and Mayes, T., 1998, HACCP implementation: a practical guide to the implementation of the HACCP plan, *Food Control*, Vol. 9 No 2-3, pp. 103-109.

83. WHO, 1993, Training Considerations for the Application of the Hazard Analysis Critical Control Points System to Food Processing and Manufacturing, WHO Document, WHO/FNU /FOS/93.3, World Health Organization, Division of Food and Nutrition, Geneva.

84. Nguyen, T., Wilcock, A. and Aung, M. (2004), Food safety and quality systems in Canada, International Journal of Quality and Reliability Management, Vol. 21 No 6, pp. 655-671.

85. Domenech, E., Escriche, I., and Martorell, S., 2008, Assessing the effectiveness of critical control points to guarantee food safety, *Food Control*, Vol. 19 No 6, pp. 557-565.

86. Loc, V.T.T., 2006, Seafood Supply Chain Quality Management: The Shrimp Supply Chain Quality Improvement Perspective of Seafood Companies in

the Mekong Delta, Vietnam, Thesis, Rijksuniversiteit Groningen.

87. Marnellos, G. and Tsiotras, G., 1999, Hazard Analysis Critical Control Point (HACCP): Implementation in Greek Industry, *Quality and Reliability Engineering International*, Vol. 15, No 4, pp. 385-396.

88. Griffith, C., 2000, HACCP and the management of healthcare associated infections, *International Journal of Health Care Quality Assurance*, Vol. 19 No 4, pp. 351-360.

89. Seward, S., 2000, Application of HACCP in the foodservice, *Irish Journal of Agriculture and Food Research*, Vol. 39 No 2, pp. 221–227.

90. Varzakas, T. and Arvanitoyannis, I., 2008, Application of ISO22000 and comparison to HACCP for processing of ready to eat vegetables: Part I, *International Journal of Food Science and Technology*, Vol. 43, pp. 1729–1741.

91. Cao, K., Maurer, O., Scrimgeour, F. and Chris D., 2004, The economics of HACCP (Hazard Analysis and Critical control point): A literature review, Agribusiness Perspectives Papers.

92. The Traceability Blue Book - Using Traceability in the Supply Chain to Meet Consumer Safety Expectations, March 2004, Accessed through: http://www.gs1belu.org/sites/default/files/publications/files/2004_traceability_blue_book.pdf.

93. Department for Environment, Food and Rural Affairs (DEFRA), 2008, Food Statistics Pocket Book 2008, London: DEFRA.

94. Buzby, J.C. and Roberts, T., 1997, Economic costs and trade impacts of microbial foodborne illness. World Health Stat. Q. 50(1-2): 57-66.

95. Mead, P.S., Slutsker, L., Dietz, V., McCaig, L.F., Bresee, J.S., Shapiro, C., Griffin, P.M., and Tauxe, R.V., 1999, Food-related illness and death in the United States. Emerg. Infect. Dis. 5: 607-625.

96. WHO, 2008, WHO Initiative to Estimate the Global Burden of Foodborne Diseases: A summary document

97. USDA, 1996, Jean, C. B., Roberts, T., Lin, C. T. J. and MacDonald, J., Bacterial Foodborne Disease: Medical Costs and Productivity Losses. Agricultural Economic Report No. (AER-741) 93 pp, August 1996.

98. Pillay V., and Muliyil V., 2005, ISO 22000 Food Safety Management Systems – The One Universal Food Safety Management System Standard That Works Across All Others, SGS Systems and Certifications Services, Surrey.

99. Tajkarimi M., 2007, New Food Safety Management Systems; ISO 22000. Accessed through http:// www.vetmed.ucdavis.edu/PHR/PHR450/2007/45 007C8T.pdf.

100. Færgemand, J., 2005, 'Standard for Food Safety Management', New Food issue 5, 2005: Accessed through http://www.bureauveritashk.com/ufiles/ ISO_22000_New _Food.pdf

101. (a) – ISO 22000:2005, (b) – ISO 22000:2018, Food Safety Management Requirements for and Organization in the Food chain, (a – 2005) first edition and (b – 2018) first revision.

102. Talbot, V., 2007, ISO 22000 Standard: A Food Safety Management System, January/March, Page 40 – 43, Accessed Through http://www.spc.int/DigitalLib rary/Doc/FAME/InfoBull/FishNews/120/FishNe ws120_40_Talbot.pdf

103. Chambers, A. F., 2007, ISO 22000 Food Safety Management Systems: An easy-to-use checklist for small business; Are you ready?, Geneva: International Trade Centre.

104. Smith, M.T., 2002, Understanding and Implementing ISO 9001:2000, Cayman Business Systems, USA. Accessed through:www.elsmar.com.

105. ISO 3720:2011, Black tea - Definition and basic requirements.

106. Product Certification Scheme for Tea (PCST), http://www.slsi.lk/web/index.php?option=com_content&view=article&id=107&Itemid=129&lang=si

107. Sri Lanka Tea Board Standards/ Guidelines for Tea, 2010, Circular No.: AL/MQS-Rev/2010, Sri Lankan Origin Teas and Other Origin Teas, Sri Lanka Tea Board.

108. FDA - US Food and Drug Administration, 2004, GMPs - Section One: Current Food Good Manufacturing Practices, Accessed Through: http://www.fda.gov/Food/GuidanceRegulation/CGMP/ucm110907.htm

109. Barendsz, A.W., 1998, 'Food safety and total quality management', Food Control. Vol. 9, pp. 163-170.

110. Dunkelberger., Edward., 1995, 'The statutory basis for the FDA's food safety assurance programs: From GMP, to emergency permit control, to HACCP', Food and Drug Law Journal Vol.50, pp. 357-383.

111. Damman, J., 1999, 'CBL-akkoord over BRC-standaard moet aantal audits reduceren', Voedingsmid delen technologie, 32: 15 17.

112. Codex Alimentarius, 2003, 'General Principles of Food Hygiene', CAC/RCP 1-1969, Rev. 4-2003

113.Surak, J. G., 2008, 'Comparison of ISO 9001 and ISO 22000'. Accessed through: http://foodsqm.files. wordpress.com/2007/11/comparison_of_iso_9001 _and_iso_22000.pdf.

114.Frost, R., 2008, 'ISO 22000 is first in family of food safety management system standards'.

115.DeMan., and John, M., 1999, 'Principles of Food Chemistry' Springer.

116.Good Manufacturing Practice for the Plastic Food Packaging Supply Chain, 2012, The Society of the Plastics Industry, Inc. (SPI) Food, Drug, and Cosmetic Packaging Materials Committee. Accessed through: http:// www.plasticsindustry.org/files/ about/fdcpmc/fdcpmc _manuf_practice%20guidelines20120126.pdf.

117.WHO., 2006, 'Basic Training Modules on Good Manufacturing Practices (GMP) - Basic Principles of GMP: Module 1 (Part 02)': Quality Management. http://apps.who.int/medicinedocs/en/d/Js14030 e/#Js14030e

118.Higgins, Kevin T., 2002, The Culture of Clean, Dairy Foods, November.

119.WHO., 2014, 'good manufacturing practices for pharmaceutical products: main principles', WHO Technical Report Series No. 986, Accessed Through: http://www.who.int/medicines/areas/quality_saf ety/quality_assurance/TRS986annex2.pdf

120.Motarjemi Y., Van Schothorst M., and Kaferstein F., 2001, Future challenges in global harmonization of food safety legislation, Food Control, 12(6): 339-346.

121.Mensah L.D., Julien D., 2011, Implementation of food safety management systems in the UK, Food Control, 22 (8): 1216-1225 doi: 10.1016/j.foodcont.

2011.01.021
http://dx.doi.org/10.1016/j.foodcont.2011.01.021

122. Jayasinghe Mudalige U., and Henson S., 2007, Identifying economic incentives for Canadian red meat and poultry processing enterprises to adopt enhanced food safety controls, Food Control, 18 (11): 1363 - 1371

123. Khatri Y., and Collins R., 2007, Impact and status of HACCP in the Australian meat industry, British Food Journal, 109 (5): 343 - 354

124. Henson S., and Hooker N.H., 2001, Private sector management of food safety: public regulation and the role of private controls, International Food and Agribusiness Management Review, 4: 7 – 17

125. Henson S., and Mitullah W., 2004, Kenya Exports of Nile Perch; Impact of Food Safety Standards on an Export-oriented supply chain.

126. Taylor E., 2001, HACCP in small companies: benefits or burden? Food Control, 12: 217-222.

127. Romano D., Cavicchi A., Rocchi B., and Stephani G., 2004, Costs and benefits of compliance for HACCP in Italian meat and dairy sectors, http://ageconsearch.umn.edu/bitstream/24983/1/sp04ro02.pdf

128. Bilalis D., Stathis I., Konstantas A., and Pasiali S., 2009, Comparison between HACCP and ISO 22000 in Greek organic food sector, Journal of Food, Agriculture and Environment, 7(2): 237-242

129. Henson S., and Holt G., 2000, Exploring incentives for adoption of food safety controls: HACCP implementation in UK dairy sector, Review of Agricultural Economics, 22(12): 407-420.

130. Turner C. R., Ortmann G., and Lyne M., 2000, Adoption of ISO 9000 Quality assurance standards

by South African agribusiness firms. Agribusiness, 16(3): 295-307.

131. Maldonado-Siman, E., Ruiz-Flores, A., Nunez-Dominguez, R., Gonzalez-Alcorta, M., and Hernandez-Rodriguez B. A., 2009, Level of adoption of quality management systems in Mexican pork industry, In D. Li and Z. Chunjiang (Eds.), Computer and Computing Technologies in Agriculture II, Boston: Springer, 3: 1747-1756.

132. Deohar S.Y., 2003, Motivation for and cost of HACCP in Indian food processing industry, Ahmadabad, India: Indian Institute Management Department of Census and Statistics, 2012, Sri Lanka.

133. Thompson V. A., 1965, Bureaucracy and innovation, Administrative Science Quarterly, 10(June): 1-20.

134. Daft R., and Becker S., 1978, Innovation in organization: innovation adoption in school organizations, New York, Elsevier.

135. Mazzoco, M. A., 1996, HACCP as a business management tool, American Journal of Agricultural Economics, 78(3): 770-774.

136. Henson S., Holt G., and Northern J., 1999, Costs and benefits of implementing HACCP in the UK dairy processing sector, Food Control, 10(2): 99-106.

137. Lokunarangodage C.V.K., Wickramasinghe I., and Ranaweera K.K.D.S., 2015, Review of ISO 22000:2005, Structural synchronization and ability to deliver food safety with suggestions for improvements, Journal of Tea Science Research, 5(12), 1-12 doi: 10.5376/jtsr.2016.06. 0002

138. Herath D., Hassan Z., and Henson S., 2007, Adoption of food safety and quality controls. Do firm characteristics matter? Evidence from the

Canadian food processing sector, Canadian Journal of Agricultural Economics, 55(3): 299-314.

139. ISO 22000:2005, Food Safety Management Requirements for and Organization in the Food chain, 2005, First edition.

140. Hirata R., 2001, Implementation of 5S in Mexico. Retrieved July 4, 2012, from http://www.keisen. com/documentos/Implementation%205S-in-Mexico.PDF.

141. Osada T., 1991, The 5S's: five keys to a total quality environment. Minato-ku: Asian Productivity Organization.

142. Kumar., Sudhahar., and Dickson., 2007, Performance Analysis of 5-S Teams Using Quality Circle Financial Accounting System. The TQM Magazine, 483-496

143. Daud Khairul and Wan Rosmanira., 2006, An Empirical Study on the Effects of Service Quality Towards Organizational Performance in Malaysian Local Authorities. Accessed through http://www.jgbm.org /page/8%20Talib.pdf

144. Hough R., 2008, 5S Implementation methodology, Management Services, 35(5), 44-45

145. Van Patten., 2006, A second look at 5S. Quality Progress, 39/10, pp. 55-59

146. Howell V.W., 2009, 5S for success, Ceramic IndusflY, August-September

147. Samuels G., 2009, 5S (Sort/Set/Shine/Standardize /Sustainability). Converting Magazine, 27/12, pp. 25-26

148. Cooper K., Keif M.G., and Macro K.L., 2007. Lean printing: pathway to success. PIAIGATF Press, Sewickly, PA.

149. Maggie L.Y., 2006, Library as place: implementation of 5S system. Journal of East Asian Libraries, 139, 57-67

150. Goetsch D.L., and Davis S., 2010, Quality Management for Organizational Excellence: Introduction to Total Quality. United States of America: Pearson, Prentice Hall

151. Wakhlu B., 2007, Total Quality: Excellence through Organization-Wide Transformation, New Delhi: S. Chand and Company Ltd.

152. Janakiraman R., and Gopal R.K., 2007, Total Quality Management: Text and Cases. New Delhi: Prentice-Hall of India.

153. Parrie J., 2007, Minimize Waste With The 5S System. Spring, 30-35

154. Gapp R., Fisher R., and Kobayashi K., 2008, Implementing 5S within a Japanese Context: An Integrated Management System. Management Decision, 565-579, http://dx.doi.org/10.1108/00251740810865067

155. Salaheldin I., 2009, Critical Success Factors for TQM Implementation and Their Impact on Performance of SMEs. International Journal of Productivity and Performance Management, 58/03, 215-237, http://dx.doi.org/10.1108/17410400910938832.

156. Helena L., Christina S. and Annika O.: Traceability in Food Supply Chain: Towards Synchronized Supply Chain. : Lund University Publications, 2008 September. Available at http://lup.lub.lu.se/record/1165966

157. Stadig, M., Breg, B., Bergström, B., Janson, C.-G., Karlsson, R., Wiik, L., and Johnsson, M. 2002, Spårbarhet i Livsmedelskedjan, SIK, SIK-Dokument 161.

158. FAO Traceability, Supply Chains and Smallholders: Case-Studies from India and Indonesia. *Committee on Commodity Problems. Intergovernmental Group on Tea* Seventeenth Session, 2006, p 1 – 21, Accessed on ftp://ftp.fao.org/docrep/fao/meeting/011/j8316e.pdf

159. Hand Book 7.1: *Traceability and Labelling*, Available at: http://pip.coleacp.org/files/documents/edes/publications/EDES%20fascicule%207-1_EN_web.pdf. [Accessed 04th January 2014]

160. ISO 9001:2000 standard

161. Popper, D.E., 2007, Traceability: Tracking and privacy in the food system, *Journal of Geographical Review*, Vol. 97, No.3, p. 365-389.

162. UNEP. "Green Economy and Trade": Green Economy Report's Agriculture chapter 2. Available: http:// www.unep.org/greeneconomy/Portals/88/GETReport/pdf/Chapitre%202%20Agriculture.pdf, 2013. p 45 – 88.

163. Kelepouris, T., Pramatari, K., & Doukidis, G. "RFID-enabled traceability in the food supply chain", *Industrial Management + Data Systems,* vol. 107, no. 2, 2007. p. 183.

164. Morrison, C. "Traceability in food processing: an introduction," in Food authenticity and traceability, M. Lees, ed., Woodhead publishing limited, Cambridge, 2003. p. 459-471.

165. Van Dorp, K.-J. "Beef labeling: The emergence of transparency", *Supply Chain Management,* vol. 8, no. 1, 2003. p. 32-40.

166. Viaene, J. & Verbeke, W. "Traceability as a key instrument towards supply chain and quality management in the Belgian poultry meat chain", *Supply Chain Management,* vol. 3, no. 3, 1998 p. 139.

167. Wang, X. & Li, D. "Value Added on Food Traceability: a Supply Chain Management Approach", *2006 IEEE International Conference on Service Operations and Logistics, and Informatics* 2006. p. 493-498.

168. Galimberti A., Mattia F D., Losa A., Bruni I., Federici S., Casiraghi M., Martellos S, Labra M.: "DNA barcoding as a new tool for food traceability".: *Food Research International* 50 (2013) 55–63.: Elsevier Publications available at http://www.urbanbar codeproject.org/ images/pdf/SubstitutionInSea food.pdf

169. Kevin Bonsor and Wesley Fenlon, How RFID Works: Accessed through: http://electronics .howstuffworks.com/gadgets/high-tech-gadgets/ rfid3.htm.

www.ingramcontent.com/pod-product-compliance
Lightning Source LLC
Chambersburg PA
CBHW071537200326
41519CB00021BB/6517